THE NATIONAL ACADEMIES
KECK FUTURES INITIATIVE

THE INFORMED BRAIN IN A DIGITAL WORLD

INTERDISCIPLINARY RESEARCH TEAM SUMMARIES

Conference
Arnold and Mabel Beckman Center
Irvine, California
November 15-17, 2012

THE NATIONAL ACADEMIES PRESS
Washington, D.C.
www.nap.edu

THE NATIONAL ACADEMIES PRESS   500 Fifth Street, NW   Washington, DC 20001

NOTICE: The Interdisciplinary Research (IDR) team summaries in this publication are based on IDR team discussions during the National Academies Keck *Futures Initiative* Conference on the Informed Brain in a Digital World held at the Arnold and Mabel Beckman Center in Irvine, California, November 15-17, 2012. The discussions in these groups were summarized by the authors and reviewed by the members of each IDR team. Any opinions, findings, conclusions, or recommendations expressed in this publication are those of the IDR teams and do not necessarily reflect the view of the organizations or agencies that provided support for this project. For more information on the National Academies Keck *Futures Initiative* visit www.keckfutures.org.

Funding for the activity that led to this publication was provided by the W.M. Keck Foundation. Based in Los Angeles, the W. M. Keck Foundation was established in 1954 by the late W.M. Keck, founder of the Superior Oil Company. In recent years, the Foundation has focused on Science and Engineering Research; Medical Research; Undergraduate Education; and Southern California. Each grant program invests in people and programs that are making a difference in the quality of life, now and for the future. For more information visit www.wmkeck.org.

International Standard Book Number-13: 978-0-309-26888-2
International Standard Book Number-10: 0-309-26888-5

Additional copies of this report are available from the National Academies Press, 500 Fifth Street, NW, Keck 360, Washington, DC 20055; (800) 624-6242 or (202) 334-3313; http://www.nap.edu.

Copyright 2013 by the National Academy of Sciences. All rights reserved.

Printed in the United States of America

THE NATIONAL ACADEMIES
Advisers to the Nation on Science, Engineering, and Medicine

The **National Academy of Sciences** is a private, nonprofit, self-perpetuating society of distinguished scholars engaged in scientific and engineering research, dedicated to the furtherance of science and technology and to their use for the general welfare. Upon the authority of the charter granted to it by the Congress in 1863, the Academy has a mandate that requires it to advise the federal government on scientific and technical matters. Dr. Ralph J. Cicerone is president of the National Academy of Sciences.

The **National Academy of Engineering** was established in 1964, under the charter of the National Academy of Sciences, as a parallel organization of outstanding engineers. It is autonomous in its administration and in the selection of its members, sharing with the National Academy of Sciences the responsibility for advising the federal government. The National Academy of Engineering also sponsors engineering programs aimed at meeting national needs, encourages education and research, and recognizes the superior achievements of engineers. Dr. Charles M. Vest is president of the National Academy of Engineering.

The **Institute of Medicine** was established in 1970 by the National Academy of Sciences to secure the services of eminent members of appropriate professions in the examination of policy matters pertaining to the health of the public. The Institute acts under the responsibility given to the National Academy of Sciences by its congressional charter to be an adviser to the federal government and, upon its own initiative, to identify issues of medical care, research, and education. Dr. Harvey V. Fineberg is president of the Institute of Medicine.

The **National Research Council** was organized by the National Academy of Sciences in 1916 to associate the broad community of science and technology with the Academy's purposes of furthering knowledge and advising the federal government. Functioning in accordance with general policies determined by the Academy, the Council has become the principal operating agency of both the National Academy of Sciences and the National Academy of Engineering in providing services to the government, the public, and the scientific and engineering communities. The Council is administered jointly by both Academies and the Institute of Medicine. Dr. Ralph J. Cicerone and Dr. Charles Vest are chair and vice chair, respectively, of the National Research Council.

www.national-academies.org

THE NATIONAL ACADEMIES KECK *FUTURES INITIATIVE* INFORMED BRAIN STEERING COMMITTEE

MICHAEL S. GAZZANIGA, Chair (NAS/IOM), Director, The Sage Center for the Study of the Mind, University of California, Santa Barbara
C. GORDON BELL (NAS/NAE), Principal Researcher, Microsoft Research
FLOYD E. BLOOM (NAS/IOM), Professor Emeritus, Molecular and Integrative Neuroscience Department, The Scripps Research Institute
APOSTOLOS GEORGOPOULOS (IOM), Regents Professor, McKnight Presidential Chair in Cognitive Neuroscience, American Legion Brain Sciences Chair, Professor of Neuroscience, Neurology and Psychiatry, Brain Sciences Center, Veterans Affairs Medical Center
CHARLES D. GILBERT (NAS), Arthur and Janet Ross Professor, Laboratory of Neurobiology, The Rockefeller University
TODD F. HEATHERTON, Lincoln Filene Professor in Human Relations, Dartmouth College
MICHAEL A. KELLER, Ida M. Green University Librarian, Director of Academic Information Resources, Stanford University Cecil H. Green Library
GLORIA MARK, Professor, Department of Informatics, Interactive and Collaborative Technologies, Donald Bren School of Information and Computer Sciences, University of California, Irvine
RUSSELL A. POLDRACK, Director, Imaging Research Center, Professor of Psychology and Neurobiology, University of Texas at Austin
REBECCA SAXE, Assistant Professor, Department of Brain and Cognitive Sciences, Massachusetts Institute of Technology
TERRENCE J. SEJNOWSKI (NAS/NAE/IOM), Investigator, Howard Hughes Medical Institute, Francis Crick Professor, Salk Institute for Biological Studies
BRIAN A. WANDELL (NAS), Isaac and Madeline Stein Family Professor, Department of Psychology, Stanford University

Staff

KENNETH R. FULTON, Executive Director
KIMBERLY A. SUDA-BLAKE, Senior Program Director
ANNE HEBERGER MARINO, Senior Evaluation Associate
CRISTEN KELLY, Associate Program Officer
RACHEL LESINSKI, Program Associate

Consultant

BARBARA J. CULLITON, Director, NAKFI Science Writing Scholar Program

The National Academies Keck *Futures Initiative*

THE NATIONAL ACADEMIES KECK *FUTURES INITIATIVE*

The National Academies Keck *Futures Initiative* was launched in 2003 to stimulate new modes of scientific inquiry and break down the conceptual and institutional barriers to interdisciplinary research. The National Academies and the W. M. Keck Foundation believe that considerable scientific progress will be achieved by providing a counterbalance to the tendency to isolate research within academic fields. The *Futures Initiative* is designed to enable scientists from different disciplines to focus on new questions, upon which they can base entirely new research, and to encourage and reward outstanding communication between scientists as well as between the scientific enterprise and the public.

The *Futures Initiative* includes three main components:

Futures Conferences

The *Futures* Conferences bring together some of the nation's best and brightest researchers from academic, industrial, and government laboratories to explore and discover interdisciplinary connections in important areas of cutting-edge research. Each year, some 150 outstanding researchers are invited to discuss ideas related to a single cross-disciplinary theme. Participants gain not only a wider perspective but also, in many instances, new insights and techniques that might be applied in their own work. Additional pre- or post-conference meetings build on each theme to foster further communication of ideas.

Selection of each year's theme is based on assessments of where the intersection of science, engineering, and medical research has the greatest potential to spark discovery. The first conference explored *Signals, Decisions, and Meaning in Biology, Chemistry, Physics, and Engineering*. The 2004 conference focused on *Designing Nanostructures at the Interface between Biomedical and Physical Systems*. The theme of the 2005 conference was *The Genomic Revolution: Implications for Treatment and Control of Infectious Disease*. In 2006 the conference focused on *Smart Prosthetics: Exploring Assistive Devices for the Body and Mind*. In 2007 the conference explored *The Future of Human Healthspan: Demography, Evolution, Medicine, and Bioengineering*. In 2008 the conference focused on *Complex Systems*. The 2009 conference explored *Synthetic Biology: Building on Nature's Inspiration*. The 2010 conference focused on *Seeing the Future with Imaging Science*. The 2011 conference focused on *Ecosystem Services*. The 2012 conference focused on *The Informed Brain in a Digital World* and the 2013 conference will explore advanced nuclear technologies.

Futures Grants

The *Futures* Grants provide seed funding to *Futures* Conference participants, on a competitive basis, to enable them to pursue important new ideas and connections stimulated by the conferences. These grants fill a critical missing link between bold new ideas and major federal funding programs, which do not currently offer seed grants in new areas that are considered risky or exotic. These grants enable researchers to start developing a line of inquiry by supporting the recruitment of students and postdoctoral fellows, the purchase of equipment, and the acquisition of preliminary data—which in turn can position the researchers to compete for larger awards from other public and private sources.

NAKFI Communications

The Communication Awards are designed to recognize, promote, and encourage effective communication of science, engineering, medicine, and/or interdisciplinary work within and beyond the scientific community. Each year the *Futures Initiative* awards $20,000 in prizes to those who have advanced the public's understanding and appreciation of science, engineering, and/or medicine. The awards are given in four categories: books, film/radio/TV, magazine/newspaper, and online. The winners are honored during a ceremony in the fall in Washington, DC.

NAKFI cultivates science writers of the future by inviting graduate students from science writing programs across the country to attend the conference and develop IDR team discussion summaries and a conference overview for publication in this book. Students are selected by the department director or designee, and prepare for the conference by reviewing the webcast tutorials and suggested reading, and selecting an IDR team in which they would like to participate. Students then work with NAKFI's science writing student mentor to finalize their reports following the conferences.

Facilitating Interdisciplinary Research Study

During the first 18 months of the Keck *Futures Initiative*, the Academies undertook a study on facilitating interdisciplinary research. The study examined the current scope of interdisciplinary efforts and provided recommendations as to how such research can be facilitated by funding organizations and academic institutions. *Facilitating Interdisciplinary Research* (2005) is available from the National Academies Press (www.nap.edu) in print and free PDF versions.

About the National Academies

The National Academies comprise the National Academy of Sciences, the National Academy of Engineering, the Institute of Medicine, and the National Research Council, which perform an unparalleled public service by bringing together experts in all areas of science and technology, who serve as volunteers to address critical national issues and offer unbiased advice to the federal government and the public. For more information, visit www.nationalacademies.org.

About the W. M. Keck Foundation

Based in Los Angeles, the W. M. Keck Foundation was established in 1954 by the late W. M. Keck, founder of the Superior Oil Company. The Foundation's grant making is focused primarily on pioneering efforts in the areas of Science and Engineering Research; Medical Research; Undergraduate Education; and Southern California. Each grant program invests in people and programs that are making a difference in the quality of life, now and in the future. For more information, visit www.wmkeck.org.

National Academies Keck *Futures Initiative*
100 Academy, 2nd Floor
Irvine, CA 92617
949-721-2270 (Phone)
949-721-2216 (Fax)
www.keckfutures.org

Preface

At the National Academies Keck *Futures Initiative* Conference on The Informed Brain in a Digital World, participants were divided into fourteen interdisciplinary research teams. The teams spent nine hours over two days exploring diverse challenges at the interface of science, engineering, and medicine. The composition of the teams was intentionally diverse, to encourage the generation of new approaches by combining a range of different types of contributions. The teams included researchers from science, engineering, and medicine, as well as representatives from private and public funding agencies, universities, businesses, journals, and the science media. Researchers represented a wide range of experience—from postdoc to those well established in their careers—from a variety of disciplines that included science and engineering, medicine, physics, biology, economics, and behavioral science.

The teams needed to address the challenge of communicating and working together from a diversity of expertise and perspectives as they attempted to solve a complicated, interdisciplinary problem in a relatively short time. Each team decided on its own structure and approach to tackle the problem. Some teams decided to refine or redefine their problems based on their experience.

Each team presented two brief reports to all participants: (1) an interim report on Friday to debrief on how things were going, along with any special requests; and (2) a final briefing on Saturday, when each team:

- Provided a concise statement of the problem;
- Outlined a structure for its solution;

- Identified the most important gaps in science and technology and recommended research areas needed to attack the problem; and
- Indicated the benefits to society if the problem could be solved.

Each IDR team included a graduate student in a university science writing program. Based on the team interaction and the final briefings, the students wrote the following summaries, which were reviewed by the team members. These summaries describe the problem and outline the approach taken, including what research needs to be done to understand the fundamental science behind the challenge, the proposed plan for engineering the application, the reasoning that went into it, and the benefits to society of the problem solution. Due to the popularity of some topics, two or three teams were assigned to explore the subjects.

Six podcasts were launched throughout the summer to help bridge the gaps in terminology used by the various disciplines. Participants were encouraged to listen to all of the podcasts prior to the November conference.

Contents

Conference Summary 1

IDR TEAM SUMMARIES

Team 1: Develop innovative curricula that will help students develop expertise in dealing with the information overload they will encounter during and after their schooling.
    IDR Team Summary, Group A, 11
    IDR Team Summary, Group B, 15

Team 2: Develop methods to efficiently design and measure the efficacy of Internet teaching technologies. 21

Team 3: Define the trajectory, value, and risk of Extreme Lifelogging when nearly everything about a person is in Cyberspace. 29

Team 4: Indentify the ways in which the Internet positively and negatively impacts social behavior. 39
    IDR Team Summary, Group A, 41
    IDR Team Summary, Group B, 46

Team 5: Develop a new approach to assess the differences in cognitive and brain function between the brains of digital natives and digital immigrants.   51
    IDR Team Summary, Group A, 53
    IDR Team Summary, Group B, 59
    IDR Team Summary, Group C, 63

Team 6: Determine how the effects of the digital age will improve health and wellness.   69
    IDR Team Summary, Group A, 73
    IDR Team Summary, Group B, 77

Team 7: What are the limits of the Brain-Computer Interface (BCI) and how can we create reliable systems based on this connection?   81
    IDR Team Summary, Group A, 82
    IDR Team Summary, Group B, 87
    IDR Team Summary, Group C, 92

APPENDIXES

List of Podcast Tutorials	99
Agenda	101
Participants	105

To listen to the podcasts or view the conference presentations, please visit our website at www.keckfutures.org.

Conference Summary

Kirk McAlpin, Freelance Science Writer

Digital media provide humans with more access to information than ever before—a computer, tablet, or smartphone can all be used to access data online and users frequently have more than one device. However, as humans continue to venture into the digital frontier, it remains to be known whether access to seemingly unlimited information is actually helping us learn and solve complex problems, or ultimately creating more difficulty and confusion for individuals and societies by offering content overload that is not always meaningful.

Throughout history, technology has changed the way humans interact with the world. Improvements in tools, language, industrial machines, and now digital information technology have shaped our minds and societies. There has always been access to more information than humans can handle, but the difference now lies in the ubiquity of the Internet and digital technology, and the incredible speed with which anyone with a computer can access and participate in seemingly infinite information exchange. Humans now live in a world where mobile digital technology is everywhere, from the classroom and the doctor's office to public transportation and even the dinner table. This paradigm shift in technology comes with tremendous benefits and risks. Interdisciplinary Research (IDR) Teams at the 2012 National Academies Keck *Futures Initiative* Conference on The Informed Brain in the Digital World explored common rewards and dangers to humans among various fields that are being greatly impacted by the Internet and the rapid evolution of digital technology.

Keynote speaker Clifford Nass of Stanford University opened the dialogue by offering insight into what we already know about how the "information overload" of the digital world may be affecting our brains. Nass presented the idea of the "media budget," which states that when a new media emerges, it takes time away from other media in a daily time budget. When additional media appear and there is no time left in a person's daily media budget, people begin to "double book" media time. Personal computers, tablets, and smartphones make it easy to use several media simultaneously, and according to Nass, this double-booking of media can result in chronic multitasking, which effects how people store and manage memory. Although current fast-paced work and learning environments often encourage multitasking, research shows that such multitasking is inefficient, decreases productivity, and may hinder cognitive function.

MULTITASKING

The topic of multitasking and its effects resulted in a wide range of IDR Team discussions at the 2012 NAKFI Conference, from behavior and education to cutting edge technologies like the Brain-Computer Interface (BCI).

Interdisciplinary conference teams had the opportunity to imagine innovative ways in which the human brain may interact with computers in the future. Three IDR Teams (7A, B, and C) explored the BCI, which refers to direct communication between the brain and an external digital device such as a computer. While there was agreement across many groups that technology is far from "mind reading," it is already possible for the brain to directly communicate with computers in some capacity. In the relatively near future, the teams thinking about this interface imagined a scenario in which a brain-computer feedback loop will be completed, meaning the brain will not only be able to influence digital signal pathways, but computers will be able to return some input to the brain that could influence and potentially even control behavior. The teams termed this a "closed loop" brain-computer interface.

Inspired by the predominant conference topics of multitasking and attention deficit, IDR Team 7B imagined a closed-loop BCI device that could aid productivity and manage multitasking by detecting when a person's attention to one task is waning, and signaling that it would be a good time to switch to a different task better suited to the state of mind the person was in at the time. Team 7A created an imaginary company called Brain Buddy, Inc., a technological way to assist people in maintaining focus by alerting the brain when it is optimized to complete specific tasks. Another Brain Buddy product could provide therapy to people with brain disorders such as PTSD and anxiety by helping them become aware of and avoid stimuli in the environment that are known to negatively influence their disorder.

LIFELOGGING

One of the more intriguing—though strongly debated—topics at the conference was "extreme lifelogging," which is enabled by the proliferation of portable digital technologies such as headset cameras, GPS equipment, and body monitoring aids. By lifelogging, a person can record virtually everything he or she does, day in and day out, and store resulting data on the Internet.

IDR Team 3 explored the benefits and risks of having one's entire life online. Because of the potential threats to privacy and control over data posted online, the team proposed that to move forward with lifelogging, there must first be a type of "Consumer Bill of Rights" to protect identity and prevent discrimination from groups such as insurers, marketers, and employers. The team also acknowledged a technical need for an online platform with software that could begin to aggregate user-submitted data to identify which data are important. Acknowledging the current limitations of the technology and risk of posting every detail of one's life, Team 3 believed that lifelogging could potentially benefit individuals seeking to better understand the consequences of their actions and improve their health, as well as provide important data to governments and businesses seeking to understand trends in human behavior.

PERSONAL HEALTH

Researchers tackling IDR Challenge 2 were asked to determine the effect of the digital age on health and wellness. Both groups explored the possibility of creating application software (apps) aimed at improving relationships between patients and physicians, and enabling patients to be more engaged in their own health care by creating a comprehensive picture of their health history. While the medical field is rapidly advancing in technology, research and new therapies, it is difficult for patients and doctors to keep up with everything. Apps proposed by the groups would consolidate patient data from different sources for better health care, and potentially help diagnose and treat current health problems in an individual. Patients could also use health apps to monitor fluctuations in indicators such as blood pressure and cholesterol.

LEARNING ONLINE

Easy access to the Internet and digital technology has radically changed how humans learn in the modern world. The World Wide Web has brought a universe of knowledge into the classroom; enabling teachers and students to use computers as a learning tool brings unprecedented information access to every subject. Now, some classes are conducted entirely online.

To assess the changing learning landscape, two teams were asked to, "develop innovative curricula that will help students develop expertise in dealing with the information overload they will encounter during and after school." As technology changes, institutions from schools to universities will need to develop strategies aimed at training students and lifelong learners to be able to manage the massive amounts of data that they are now exposed to. Team 1B came up with the idea of a Life Long Learning Locker (L4), which could manage, filter, and adapt information based on one's personal education history and future goals. In an age of "information overload" and specialization, a system like L4 would help individuals cater to their personalized educational needs.

After designing and implementing Internet-based learning curricula, there will have to be methods to assess the efficacy of those programs. IDR Team 2 decided that current measurement tools are not sufficient; there is an acknowledged lack of data about student performance, without which it is difficult to plan future coursework and assess what programs work. Key to meeting the challenges of developing tools to measure digital-based learning criteria is gaining knowledge of diverse stakeholders and developing a framework for assessing outcomes for different technologies. In order to be able to make future recommendations, the group proposed research including determining appropriate learning technologies for different individuals, assessing what is a good mix of technology and traditional social interaction ("face-to-face" time), and learning to anticipate the effect of new technology. The team also proposed the creation of a public database that would allow users to see how the programs are evaluated by accredited organizations.

DIGITAL TECHNOLOGY AND BEHAVIOR

The effect of the new technologies on human behavior was also debated at the conference, and there was a general assumption that online communication has the potential to negatively alter person-to-person relationships. While the Internet allows people from diverse backgrounds and countries to connect in ways that would have seemed impossible a decade ago, some researchers say that purely online interactions could diminish the quality and value of human relationships, leading to impaired emotional and social development in children as well as adolescents and adults.

Team 4A proposed research on how the Internet affects strong human ties, which are relationships characterized as reciprocal, high intensity, and long lasting. People rely on strong ties for support and emotional development, and strong ties are associated with mental and physical well-being. Although online platforms such as Facebook, Twitter, and YouTube allow people to maintain many social ties with less effort, Team 4A observed that these ties may often be characterized as weak. The team hypothesized that the creation of increasing amounts of weak ties may come at the expense of important strong ties. Because the phenomenon is not well understood, the group proposed the creation of a scale they termed an Emotional (Intelligence) Quota (EQ), which could be used to measure peoples' emotional responses to interactions that form and take place on the Internet.

Team 4B approached the effect of the Internet on behavior by considering the effect of echo chambers, which result when people seeking information on the Internet are filtered toward information that is more agreeable to them or websites that are more likely to contain information that fits their opinions, rather than lead them to sources that may challenge their ideas. Internet filters exist on many different levels, from the creation of the content, to technical filters that use search engines and social network trends to rank searched material, to the individual's own preferences. Echo chambers lead to bias and polarization among people who use the Internet as a source of information. To begin to understand echo chambers and help users understand their susceptibility to bias on the Internet, Team 4B proposed research that would gauge individual awareness of bias on the Internet.

DIGITAL NATIVES AND DIGITAL IMMIGRANTS

Three IDR teams assigned were asked to assess differences in cognitive and brain function between digital natives and digital immigrants, or those who grew up using computers versus those who adopted digital technology later in life. (Team 5C pointed out that the topic is either irrelevant or truly ongoing because technology is changing so fast that today's digital native may be tomorrow's digital immigrant.) The groups did all agree that in light of an individual's different needs and experience with digital technology, the development of tools designed to optimize an individual's cognitive function in an information dense world is important.

ETHICS, LAW, AND SOCIAL POLICY

Though much of the discussion focused on advancing technology, there was also an acknowledged importance of ethics, law, and social policy. As technologies advance and become more widely available, there will be huge datasets floating around on the Internet and it will be important to determine how the data will be used and stored, and what agencies or companies will regulate this universe of information. Especially in cases of BCI and lifelogging, team members worried that information could be hacked and used against the interest of the individual collecting and using the data.

IN CONCLUSION

During the conference, IDR Teams grappled with the idea of the Internet and other digital technology as largely unexplored phenomenon in relation to neurology. Participants agreed that there was insufficient research published about the relationship between the Internet and the brain. Each topic at the 2012 NAKFI Conference seemed to imply great gains and potential dangers for humans, and the tone of the conference was that it is not clear which way the pendulum will swing.

As people move into a digital world where the possibilities seem infinite, it is important to continue to consider whether or not this is a world people will want to live in, and how individuals will maintain some degree of control over their environment. What will be the result on individual behavior and the actions of societies if we have access to unlimited knowledge, can read people's minds, remember everything that occurred in the past or know years in advance how one may die? However much humans try to harness technology and use innovation to their benefit, it may be impossible to predict the effect of technological change and anticipate its consequences. Using an interdisciplinary approach to address these challenges, IDR teams at the 2012 NAKFI Conference proposed new research to help humans understand and benefit from the digital world as we continue to evolve with it.

IDR Team Summary 1

Develop innovative curricula that will help students develop expertise in dealing with the information overload they will encounter during and after their schooling.

CHALLENGE SUMMARY

While the age in which we live has been termed the Age of Information, it also seems self-evident that the system by which we educate our youth is failing to produce the self-motivated, skilled citizen who can acquire, analyze, and create information that will contribute to the health and welfare of society. Even before the advent of the Internet and the access to enormous petabytes of machine-readable text, U.S. corporations bemoaned their requirement to spend millions of dollars to teach high school graduates to read, write, and perform basic mathematics. American students place well down the list of proficient students among lesser industrialized nations. Thanks to decades of intermittent federal investments in biomedical research, the pool of factual data amenable to analysis, and which should become part of all new physicians' operational skills is becoming so large as to be unmanageable.

This IDR Team will engage with the crisis in education and the lack of a strategy to devise tools for efficient learning and will involve the intersection of neuroscience, engineering, and medical research. Under this umbrella neuroscientists who study memory and learning, attention, and decision making, could work with engineers and educators to develop innovative curricula that would help our young students cultivate expertise in dealing with the information overload they will encounter in and after their schooling. This broad topic represents a massive opportunity to create what Branscomb, Holton, and Sonnert (2001) have termed "cutting edge research in the service of public objectives," and what is sometimes abbreviated as "Science for Society" or "Jeffersonian Science."

One major test ground for the implementation of methods for efficient lifelong learning could focus on the medical student who must learn not only the relevant facts and their application to disease mechanisms, treatment, diagnosis and prevention, but also to assimilate into that body of working knowledge all the new facts that will emerge during their careers as practicing physicians. This is also the case confronting tomorrow's clinical trainees—and the paraprofessionals who will be needed to support them; the ever more rapid medical discoveries that need to be translated into care and prevention, the lack of time to train in federally funded residency programs and additional constraints imposed on this training by maximum hour work weeks, and a national healthcare plan that will reduce the Medicare funding for post-graduate clinical training.

While the IBM-Watson device and proprietary differential diagnostic systems—costing hundreds of thousands of dollars—are beginning to enter some forms of managed health care, such computer-assisted judgments can scarcely be an acceptable form of medical practice. Therefore, the underlying problem remains of devising an educational system that will not only motivate students to become skilled in basic academics, in the technology of any occupational discipline, but also evolve into a citizen who contributes back to society. Can a formal education system include only academic basics for collecting knowledge, or should it also include understanding the value of that knowledge, the processing of knowledge, the emotional value of inspiration, creativity, risk, and resilience from failures?

Two developments based on the use of information technology to support instruction and discovery that show some promise are **learning management systems** (LMSs) and the developing **Semantic Web.** The use of **learning management systems,** both proprietary and open source, to support traditional face-to-face instruction has been in place and widely practiced for well over a decade, but there is decidedly little scientific assessment of effectiveness. For those caught on the analog side of the digital divide, persons located in places ill served by telecommunications, the cadre of essential computer and network support personnel, and the instructors adept and willing to exploit the possibilities of LMSs, the possibilities are limited.

For those in the middle-of-the-bell-curve of usage of information communications and technologies, there seems to be benefits to the use of LMSs, such as: more efficient administration of courses with more supplemental materials, collaborative document creation, online study sessions, and practice and exams with final grades transmitted directly to student information systems. On-campus users of LMSs seem to interact more with the Web-based course support than do commuter students, but both improve performance in a course supported by an LMS. Among the environmental elements not tested is that of the engagement of social networking behaviors (via the likes of Facebook, Twitter, blogging, and similar) on intellectual development. Seemingly important information appearing as a result of a Web search on Google or Bing or similar have not been studied with the possible exception of the Stanford experiment by Thrun et al., who broadcast a course on Artificial Intelligence to over 100,000 "students" anywhere on this Earth. These experiments, and others such as asynchronous audio or visual course LMSs, deserve critical analysis.

The **Semantic Web**, a theoretical proposition envisioned by Tim Berners-Lee, the "inventor" of the World Wide Web, is intended to supersede the present chaos of the Web by the creation of a massive collection of information objects on the Web that "understand" one another in a machine sense, to create a structured web of documents enabling much more efficient retrieval of relevant information objects in response to human queries. As the number of machine readable statements of relationships with associated, unchanging Web addresses for the related information objects expands dramatically, the likelihood of the improvement of discovery of numerous ideas, objects, and references in numerous formats and genres that are highly relevant increases, while the time and effort necessary to search and retrieve those will decline dramatically, and hot links to the initial investigative entry will be created. The potential for computer-assisted lifelong learning as well as computer-assisted research at the highest level is also increased, without regard for the flood of new data joining the swamp of older data on the Web. The ability of these new agents to increase our intellectual reach without the necessity of remembering any more than the essence of the most relevant documents and the taxonomy of terms in the combined essences of one's interests will expand our ability to deal with the flood and the swamp. Humans' responsibilities to remember will become more nuanced, but our abilities or duties to understand, analyze, evaluate, and then apply knowledge will increase. The creation of new knowledge and the discovery of new relationships among ideas and facts and systems will advance the state of our comprehension of our world from the most atomic or even subatomic frame to the cosmological. The contributions made possible by this quiet revolution will address matters of human health, our environment, transportation systems, education, and all the other aspects of our lives susceptible to rational thought and discourse.

Key Questions

- Can human knowledge acquisition and creation be made more efficient or more efficacious with computer-assisted learning systems, and if so, at what price, in what time, and in which arenas of society?
- In which domains of learning could such devices improve learning efficiency and in which are such improvements less certain?
- Is medicine/health the most societally-important test ground in which to apply such a learning system, or would the end result be improved with a longer time frame by starting with another test ground such as infants/toddlers?
- What is the evidence that scientific understanding has become more comprehensive and facile since scholarly journals went digital? Do scientists read more or less? Do they have deeper knowledge of their areas of specialization? Has time spent on literature researching improved the speed or breadth of discovery?
- Is there evidence that the brain is changing as the attributes of the World Wide Web, including social networking, are accessed often by various age cohorts?

Reading

Berners-Lee T, Hendler J, and Lassila O. The Semantic Web. *Scientific American* 17 May 2001;284;34-43.

Bird S, Bradshaw D, Chan WK, Clark C, Mears A, Milton U, Nuttall C, Palin A, Petroff A, Scholten B, Smith E. Online learning. *Financial Times* (Special Report) 12 March 2012.

Bloom B. Taxonomy of learning domains. 5 June 1999.

Branscomb L, Holton G, Sonnert G, Packard and Sloan Foundations. Science for society: Cutting edge basic research in the service of public objectiveness. A blueprint for intellectually bold and socially beneficial science policy, 2001.

Brown E. IBM Watson: Final Jeopardy! And the future of Watson. TED Presentation.

Council on Library and Information Resources. Linked data for libraries, museums, and archives: survey and workshop report October 2011.

Keller MA. Linked data: a way out of the information chaos and toward the semantic web. *Educause Review* July/August 2011;46(4).

Thrun S, et. al. MIT/Stanford artificial intelligence experiment.

Watson, D. Pedagogy before technology: Re-thinking the relationship between ICT and teaching. Education and Information Technologies 2001;6(4):251-266.

Because of the popularity of this topic, two groups explored this subject. Please be sure to review the other write-up, which immediately follows this one.

IDR TEAM MEMBERS—GROUP A

- Giorgio A. Ascoli, George Mason University
- Vinton G. Cerf, Google, Inc.
- Alan D.J. Cooke, University of Florida
- Carolyn Crist, University of Georgia
- Felice C. Frankel, Massachusetts Institute of Technology
- Matthew K. Henley, University of Washington
- Roy Pea, Stanford University
- Shriram Ramanathan, Harvard University
- Laura L. Symonds, Neuroscience
- Mercedes Talley, W.M. Keck Foundation

IDR TEAM SUMMARY—GROUP 1A

Carolyn Crist, NAKFI Science Writing Scholar
University of Georgia

IDR Team 1A was asked to develop innovative curricula that will help students gain expertise in dealing with the information overload they will encounter during and after their schooling. By looking at the needs related to attention, multitasking, and executive control in various age groups, this group debated whether to explore how to help increasingly distracted children who are in school, how to harness technology to create more data and information about the education environment, or how to teach students to mine and analyze the already overwhelming amount of data within various disciplines. As part of this discussion, the group agreed on the value of learning to preserve attention, focus, and ignore distracting influences.

The world now has an attention economy, and the learner has increasingly fragmented attention. Though there is more information than ever, it doesn't necessarily mean media consumers are overloaded by it, the team decided. What media users are running out of is time and attention, and the problem occurs when attention is distracted, or users simply have an inability to filter the information. E-mails, chat messages, and social media leave no time for serious work. This is happening at the same time that the demands for standards of learning, higher literacy, and specialized skills are ramping up, especially in the face of changing technology and an interdisciplinary approach to tackling problems. As part of this, the team acknowledged the expanding gap between the limited physical world of traditional schooling and the parallel virtual universe in which different and superior learning can be experienced, experimented, and designed as a vision of personalized mobile learning.

Several Solutions to Study Self

The team developed a "quantified self for learning" that would allow a user to measure, monitor, and make informed choices about his or her media consumption, time management, and productivity.

Dubbed many names—Nagster, Weight Watchers for Attention, or Attention Diary—the team created an idea for a program to help students maintain a log of what they have learned, as well as ways to capture data about themselves to analyze when and how they learn best. The idea is to foster reflection via easy access to a log of where they spend time and what they achieve.

For example, did the student eat breakfast before school, what is her emotional state while studying a certain subject, or what is the educational environment (such as witnessing a fight in the hallway) that is directly affecting her learning state? As part of this, the group proposed attention management tools that can track and visualize a user's attention allocation related to goals and targets, coach behavior to help intervene and provide attentional focus, and provide a social environment to collaborate and stay accountable with other users in a defined community. A key component would include easy and automatic logging to reduce distraction and multitasking from the tool itself.

The program would include learner goals (study for a test, learn a language, practice sports), categories of attentional expense (for each activity contributing to a goal), measures of attentional expenditures (automated and subjective), and achievement progress for goals as part of attention allocation. The interface designs for this idea are variable, but the group agreed it needs to incorporate agency and be learner-designed and generated. Certain levels for elementary school, high school, college, and senior adult users will allow for scalability, user scenarios, and applicability related to federal laws. The group expressed equity concerns when it comes to differing ages, cultures, socioeconomic statuses, and learning disabilities.

Several specific ideas include a Mobile Focus Dashboard that captures and synthesizes this data for users, which visualizes attentional activities in relation to goals. A Nagster/Navatar computer application could represent an "embodied agent" or accountability partner. Reminiscent of Pinocchio's Jiminy Cricket, Sesame Street's Oscar the Grouch, or Star Wars' Yoda, it reminds users when and how long to spend on a task, as pre-determined by the user. In addition, the team explored the idea of attention credits as a virtual economy that gives points for time and attention allocation that can be used throughout the week and help users to become aware of the limited inputs and outputs related to time. Finally, an immersive game could help users, especially young learners, to discover the effects of media multitasking as an avatar in a fast-moving world that must determine how to spend time and attention.

Overall, the aim is to help learners prepare for self-regulated inquiry, sense-making, learning design, collaboration, and self-reflection on attention, productivity, intelligence, and improvement. The idea is to create lifelong learners through a digital learning model that allows for agency and power.

Meet Marina, a Media User

As part of the design process, the team split into groups to sketch four "day-in-the-life" scenarios to determine the efficacy of their ideas, especially the Mobile Focus Dashboard. The groups addressed how elementary school, high school, college, and senior learners would use the system, and they discovered requirements for designs and issues for research to be effective at the individual, family, and institutional levels. To truly determine if the dashboard could be implemented, the group wished for more time to try rapid prototyping and piloting through iterative participatory design.

For example, as part of the high school user group, Marina is a 16-year-old junior in high school whose family immigrated from Guatemala when she was 5. Though her mother is not yet fluent in English, she is eager for Marina to develop knowledge and skills for a science and mathematics career. Marina's four brothers and sisters and grandmother live with her in San Jose, California. As part of her media use, Marina is engaged with English movies and music, as well as Spanish telenovelas and music. She has frequent video chats with her family in Guatemala and constantly updates her friends on Facebook. She realizes that doing well in school is the key to helping her family and feels a responsibility to do well, but she's caught between keeping in touch with her family and friends and earning the scholarship that could take her to college. She may not complete homework for various reasons—chatting with family and friends online at night, caring for younger brothers and sisters, and participating in afterschool clubs that will enhance her resumé. The theme is that she has multiple cultural identities with multiple implications for media use and multitasking. As part of this learning system, she needs the ability to set goals but also allow time for reflection, exercise, and time for family and friends. By investigating specific details about Marina, the group acknowledged the importance of looking at both the quantity and quality of attention, as well as balance in a learner's life.

Additional Questions for the Future

Much research is still needed regarding the concepts related to traditional and virtual education and how to manage the information learned during formal and lifelong schooling. The group developed the following questions:

Multitasking

- How are high multitaskers and low multitaskers defined, and how do they develop?
- What prevents a low multitasker/media consumer from becoming a high one?
- What do we value in relation to multitasking and media use, and do we need to help people move between categories?
- What types of multitasking scenarios are effective and ineffective?
- What measures are we using in the lab to study these multitasking scenarios, and are they truly relevant to the real world?

Attention and motivation

- How does this relate to memory and attention, and are we creating generations with attention disorders? How is this affecting work environments?
- How might social sharing and gaming elements, such as competition and prizes, help learners to manage attention?

- What are the intrinsic/extrinsic motivation variations, and how could this be incorporated with attention credits, or points awarded for time spent on a task?
- Are particular forms of distraction more disruptive than others, and how does this vary by learning domain and task?
- What is the impact of age, ethnicity, and other individual differences on distractibility?
- How can the user learn to manage his or her attention when not using the focusing-feedback system?

Data and media use

- How can researchers mine the data that is gathered from all of these sources and teach college students to harness the deluge of data to annotate new information and make discoveries?
- What are the social implications of high media use, short attention spans, and reliance on technology?
- What are the privacy and security concerns related to personalized mobile learning and collecting data?

IDR TEAM MEMBERS—GROUP B

- John-Paul Clarke, Georgia Institute of Technology
- Scott T. Grafton, University of California, Santa Barbara
- Shonali Laha, Florida International University
- Julie Linsey, Texas A&M University/Georgia Tech
- Wei Lu, University of Michigan
- Dejan Markovic, University of California, Los Angeles
- Jun Wang, Syracuse University
- Debra L. Weiner, Children's Hospital Boston
- Michelle Yeoman, Texas A&M University

IDR TEAM SUMMARY—GROUP 1B

Michelle Yeoman, NAKFI Science Writing Scholar
Texas A&M University

IDR Team 1B was asked to develop innovative curricula that will help students acquire expertise in dealing with the digital information overload they will encounter during and after their schooling. As technology changes the world, the skills needed for success evolve. Students need to be proficient at analyzing, evaluating, and synthesizing large amounts of varying information. Adults need to be life-long learners with the ability to navigate through overwhelming amounts of information in order to succeed both at work and at home. Unfortunately, the fractured nature of the information overload encourages superficiality and generalizations, and discourages deeper analysis and synthesis.

Clearly, digital technology affects how students and adults learn—just what those affects are is not entirely clear. Also unclear is how this technology will continue to change society. What will be the educational needs of students in 2030? 2050? What challenges will the students of today face at work and at home? Are there educational strategies that can better prepare students for the future?

At first, the team had trouble narrowing the scope of this topic. One member suggested that students lack discipline and that a return to more traditional and structured classroom approach is needed. Other members emphasized the value of educational games in motivating and engaging students. Another suggested that parents limit the use of technology for their children. However, the majority of the team felt that this approach is impractical and could not be widely implemented. One team member suggested the creation of a protected web space, targeted at middle school students, which would help them learn how to manage and evaluate information. After some discussion, the team decided that its first goal was to characterize the deficits in the current education system.

Education in the Digital Age

The education system does not prepare students to cope with the information overload they encounter in this complex and changing digital age. Students lack the fundamental knowledge and analytical skills to evaluate information for relevancy, accuracy, and applicability. At the same time, the amount of information on the Internet continues to increase. Our education system needs to adapt to this new reality and prepare students for the information overload they already encounter.

The team then discussed what it considers the characteristics of superior education: fostering fundamental skills, encouraging small student groups, making use of educational games, and personalizing learning experiences. With fundamental skills, students can learn to manage and filter the vast amounts of information available to them. Breaking students into smaller groups encourages interaction and fosters social development. Educational games can illustrate concepts and allow students to apply their knowledge. Catering to individual learning styles can also make learning fun, engaging, and efficient. The team agreed that these superior education methods should be incorporated into its curricula proposal.

While the team challenge was to assess the negative impacts of technology, team members also discussed the benefits. Digital technology, such as interactive learning games, can be a powerful educational tool that can improve learning, both in the classroom and at home. In order to maximize the potential benefits of the digital age on education, the negative impacts of information overload must first be resolved. One team member suggested that students and adults need a Hitchhiker's guide to the Internet, which will help them navigate through the confusing and vast digital world.

The education crisis is a vast and complicated subject. The team decided that attempting to change classroom curricula for K-12 education is beyond the scope of the challenge. As a consequence, the team decided to focus on solutions to the information overload that could be implemented outside of classrooms or at the college levels.

The team discussed two basic approaches to information overload: 1) design an adaptive interface that filters the overwhelming amount of information for the user, and 2) teach students and adults fundamental skills and knowledge which enables them to filter information themselves. With this in mind, the team decided that an interactive, adaptive, and modular online system has the potential to do both—help users filter the information overload, and teach users how to analyze and evaluate this information themselves. IDR Team 1B proposed a model for this system called the **Life-Long Learning Locker (L^4)**.

Life-Long Learning Locker (L4)

The L^4 would be an adaptive learning management system. The purpose of this system would be to select educational content based on individual learning styles and observed learning behaviors. Unlike other learning systems with a similar goal, this system will be customizable and adaptive—becoming uniquely personalized and tailored for the user. This learning system will be the optimal teaching interface, by adapting to the user's individual learning style, interests, and current skill level in that subject.

As the optimal learning interface, the L^4 will do the following:

Educate. The system will incorporate interactive tutorials, prepared educational content such as videos, and games to educate both students and adults.

Manage. L^4 will have a customizable organizational structure with icons of shelves, bins, and folders in which to manage the vast amounts of information. Users can keep past content that they find useful. An example would be a pre-medical student keeping notes from an introductory anatomy class. Ideally, the system will be used from elementary school through adulthood, so that users have access to materials throughout their lives.

Filter. The system will incorporate a powerful search engine that helps users find and filter information available on the Internet. The search engine will uniquely tailor results to suit the user's educational level, preferences, and needs. The filter would rely on ontologies (maps of related information) of semantic annotated Web pages (Web pages tagged with meaningful, identifying information).

Adapt. Search engine results and the sequence of educational content will be based on the user learning profile. The user profile will be characterized by user-supplied responses, observed characteristics of the user, and in response to observations in the community.

Potential problems with the system include privacy issues. Some people may be concerned because an outside entity is storing vast amounts of their personal information. However, current research suggests that the idea of privacy may become antiquated—some users in the Facebook generation have grown up without privacy and may not value it. Thus, privacy may be an issue for some people, but is unlikely to be an impediment to the application and use of this system.

Another potential problem with this system is motivation. However, multiple methods can be used to motivate the user. Games and interactive tutorials can make learning more enriching, engaging, and fun, which will encourage self-directed learning and increase motivation. A points system, among friends or with classmates, can also spur competition and act as a motivator. In addition, the L^4 can be implemented in traditional and distance education classrooms, and a teacher can provide motivation for using this system by assigning research topics or educational content. Some team members questioned whether adults would be motivated to use the L^4. However, other members suggested that adults who currently struggle with information overload may be motivated to use this system, provided that it is helpful and practical. Additionally, students who have used the L⁴ during their education may be likely to continue using the system through adulthood.

Conclusions

The digital landscape simultaneously unites people from around the world, and distances individuals who are sitting across each other at dinner. As with any new tool, there are benefits and trade-offs for society and the individual. People are assaulted with information that comes in multiple forms and with varying degrees of accuracy and value. Attention spans are becoming shorter as people become addicted to the constant stream of incoming, fragmented data. Multitasking has developed as a way to cope with the information overload but is appropriate only for superficial tasks. As a result, people are becoming increasingly unable to synthesize the vast amount of information that is now readily available.

The team's goal was to design a proposal that integrates available and emerging digital technologies to help students and adults manage the information overload that plagues this modern digital age. The team proposed the development of an adaptive, interactive, modular learning system that both educates and adapts to users. Ultimately, the system would help people manage and evaluate the overwhelming amount of available information by adapting itself to the user's unique learning preference, then acting as a guide through the changing digital landscape. Were this hypothetical interface to be developed, users could not only manage the overwhelming information overload, but can even excel at its management.

IDR Team Summary 2

Develop methods to efficiently design and measure the efficacy of Internet teaching technologies.

CHALLENGE SUMMARY

Over the last one-hundred years, a series of new technologies have each promised to revolutionize teaching. Slides, film and videos were all plausible technologies for replacing the classic form of instruction, in which a teacher lectures a class. Despite the many new technologies that might have changed teaching, conventional classrooms are still dominant. The World Wide Web—with the ease of posting videos and designing interactive online tests—is an attractive mélange of technologies that offer up considerable potential for influencing and improving education. Will things be different this time? How will we know the efficacy of the new Web-based teaching and learning technologies?

One reason why the Web technologies may be different is that many, many more people are authors. Pelli and Bigelow estimate that the number of authors is growing tenfold each year. The new transformation extends widespread literacy to widespread authorship. This will have an enormous impact on the range of available educational material or at least material that could be useful in educational settings.

A second reason is that Web usage, unlike slides or videos, is integrated into the typical day of many people. By integrating educational materials into daily communication and social networking tools, the new technologies have advantages that were not part of the earlier innovations.

Education is based on testing, credentialing, and building affinity groups. The Web is likely to be able to perform each of these functions, but the way in which it does so will differ from traditional classroom teaching.

How can we measure whether the new ways are effective? How can we use these measures to guide and improve the efficacy of these technologies over time? The variety of uses of Web technologies for teaching, for learning, and for research across the disciplines is impressive, but how can we assure that there is sharing and promulgation of those uses for re-use and re-mixing among the disciplines and learning environments?

Key Questions

• What can we learn from the history of introducing new technology into Education? Is there any difference with the new technologies that suggest we will have different outcomes?

• Are we prepared to measure the efficacy of the new technologies and use this ability to nudge them into useful directions?

• New software applications and new possibilities arising from them are being promoted to help authors produce educational material. Will this increase authorship and change the range of materials? What is the reasoning at academic and textbook publishers and in the software industry?

• In the US and many other Western countries, education and research are completely intertwined. What are the implications for research Universities if the education model is transformed?

• What are the implications for middle school and high school if we rely on instructional materials that are distributed, used, and take advantage of Web technologies? What is the new role for teachers and teacher education? Does it change dramatically? What is the uptake of such technologies in our public schools? Has the use of modern information technology improved teaching and learning in the K-12 grades?

Reading

Bhattacharjee Y, Clery D, Dede C, Greenfield PM, Hines DJ, Jasny BR, Mervis J, Miller G, Normile D, et. al. Special science online collection: Education and technology. *Science* (Special Issue) 2 January 2009;323(5910).

Gates Foundation. Next generation learning: The intelligent use of technology to develop innovative learning models and personalized educational pathways. 2010.

Khan S. Let's use video to reinvent education. You Tube (Video) March 2011.

Nochese F. Khan Academy: My final remarks. Action-reaction: Reflections on the dynamics of teaching 10 May 2011.

Pelli DG and Bigelow C. Nearly universal literacy is a defining characteristic of today's modern civilization; nearly universal authorship will shape tomorrow's. *Seed* Magazine 20 October 2009.

Robinson K. Schools kill creativity. TED Presentation 2006 June.
Stanford University. Stanford University EPGY.
Thompson C. How Khan Academy is changing the rules of education. *Wired Magazine* 15 July 2011.
Thrun S. University 2.0. *Livestream* (Video) 2012.
Watters A. The wrath against Khan: Why some educators are questioning Khan Academy. Hack Education 19 July 2011.
Widom J. From 100 Students to 100,000. *ACM Sigmod* 24 February 2012.

IDR TEAM SUMMARY

Zahra Hirji, NAKFI Science Writing Scholar
Massachusetts Institute for Technology

IDR Team 2 was charged with the herculean task of developing methods to efficiently design and measure the efficacy of various Internet teaching technologies. Today's teachers and self-learners have more tools at their disposal than at any previous point in history, in large part due to the proliferation of the World Wide Web. Out of all the Internet tools now available—video-audio lecture surrogates, adaptive tutorials, games, online quizzes/assessments, wikis, blogs, chat systems—how does one possibly decide which tool or tools to adopt?

To make informed decisions, teachers and self-learners need access to technology evaluation data. This information would ideally be housed in a single location and consist of standardized assessments that allow for technology comparisons. Unfortunately these data and infrastructure are largely nonexistent. As a result, users are forced to select tools based on the hyperbole of marketing rather than on scientifically rigorous assessment.

IDR Team 2 outlined the actions needed to give teachers and users pursing online education, whether formal or informal, whether for themselves or to teach others, access to the assessment information they deserve and need. The envisioned end goal also includes equipping technology users with the resources to conduct their own qualified assessments.

The Internet Is Different

The current mania surrounding Web-based teaching technologies is nothing new. Back in the 1870s, Boston schools thought the school slate—personalized chalkboards—would revolutionize education. Similar claims were then made about the film projector, radio, television, and laptop.

Educators are always looking to improve student performance, and the untapped potential of new technologies is deliciously appealing.

Schools have adopted these various technologies to different degrees, but there is little proof that such tools actually improve student performance over prolonged periods of time. As a result, the way education is generally conducted has not been dramatically altered in modern history. From K-12 to higher education, teachers and professors still commonly lecture in front of the classroom, hand out paper exams, and employ textbooks for homework.

So, is the Internet actually any different? The team definitively decided yes. Unlike previous technologies, Web-based ones can build upon the strengths of older technologies. Televisions brought pre-recorded visual material into education, but unless you had a TV and school videos at home, that technology was confined to the classroom. But using the Web, videos can be made accessible at any time—in the classroom or outside the classroom—to anyone using a computer, smartphone, or tablet.

These modern technologies also break the barrier between the classroom and world with the capacity to "inreach," or bring untapped information or people (parents, alumni, other teaching professionals) into the classroom in a virtual context. In addition to stimulating better learners, these tools may benefit teachers by automating normally exhaustive tasks, such as grading homework or tests, which then frees up the teacher's time for more constructive personal interactions with students. These tools can also help teachers track the progress or tailor workloads of individual students more readily. Finally, these technologies benefit those outside formal education too, where it is now easier to refresh old skills or pick up new ones on your own.

Is the Internet a Threat?

Internet tools are flexible, interactive, and adaptive technologies—and can be accessed from the privacy of your own living room. Moreover training teachers in new technologies is time and cost intensive. If these are so great, accessible, seemingly independent, do we even need teachers or traditional classrooms? Now this may seem like a bold question, but many state universities in the process of slashing funding are turning to massive open online class [MOOC] alternatives, such as Coursera and Open Learning Initiative, to replace, or at least supplement, traditional lectures. While this has not yet corresponded with the widespread layoffs of associate or adjunct professors, the general mood among some college campuses has been described as tense, to put it mildly.

Despite the appeal of MOOCs, there is little evidence that such technologies are just as effective, or better than traditional lecture classes. For the few subjects where evidence does exist and is relatively positive, mainly for some math and science courses, it is not yet possible to extrapolate whether this teaching format will work for other subjects, such as the social sciences and humanities.

Many of the team members are professors, and a few even had experience with MOOCs. For the MOOC users, the online portion of the class served primarily as a supplement to in-class time rather than a complete alternative. In other words, these technologies were job-helpers, not job-eaters. Regardless of the future role of MOOCs in the greater education systems, the ability to assess online technologies comprehensively is paramount. The last thing society want is the complete eradication of in-class learning for sub-par online technologies.

What Criteria, Information, and Tools Are Necessary for Assessment?

When approaching the task of building useful web teaching technology assessments, it is necessary to review what criteria, information, and tools are necessary for assessment—and how much of that material is currently known or available. The team members realized there is much we do not know or have at this point in time. In response, the team identified the three major challenges: (1) knowledge of diverse stakeholder needs, (2) a lack of accessible student performance data (at an individual or even aggregate level) for different technologies, (3) and no existing cohesive framework and infrastructure for assessing and evaluating learning outcomes for such technologies.

Assessment criteria

Ideally, Internet education technology assessments should meet the criteria of diverse stakeholders—K-12 teachers, professors, self-learners, and even employers. Some general student performance assessment points include fact learning, life skills, student engagement, student retention and completion, student diversity and access, course cost, and attitudes toward lifelong learning. Is it possible for any one assessment to actually evaluate all the criteria listed, or will separate specific assessments be more useful, practical?

Data collection

In order to assess diverse performance measures, there needs to be available data on student performance, ideally measured continually or at least routinely throughout the learning process. Many existing tools do not collect or share these data with the public. Of the ones that do both, educators can still only access very aggregated information. This is one of the major charges against the popular online software Khan Academy. Despite its popularity and plethora of testimonials, the company does not make its data accessible to the public for external evaluation.

On the flip side, the mentality that all data are good data is not necessarily true (unless you are a researcher). If educators could identify metrics for which they specifically want data, such as student quiz and assessment scores and time spent per online homework assignment, then existing Internet technologies could be reprogrammed to collect tailored information that meet stakeholder specific needs.

Assessment framework & infrastructure

There is no existing framework and infrastructure for assessing and evaluating learning outcomes for these modern technologies. In fact, there are few relevant frameworks for non-Internet based education technologies. One notable example, however, is the online resource *What Works*. This website is a database of researched and reviewed educational programs, products, practices, and policies for the K-12 school system. The website's existing information is high quality, but there is relatively little site content. Moreover, this resource does not currently account for Web-based teaching technologies.

Recommendations

The research team came up with two sets of recommendations to pursue the advancement of World Wide Web teaching technology assessment: the development of assessment infrastructure and the proposal of research topics to take advantage of this new infrastructure.

Building appropriate infrastructure

The team decided a future Internet teaching technology assessment infrastructure should be modeled after *What Works*. This resource would be a more scaled-up comprehensive version of *What Works*, including a public database of teaching technology assessments, relevant for K-12, higher education, and life-learning needs. All available assessments would be a certain high quality and have to go through some level of review. Moreover, assessments would ideally be standardized to allow for technology comparisons.

In addition, there should be automated tools available for good experiment and assessment design so that website users could contribute their own experiences and feedback.

Future research goals

Once this new database is constructed, various research studies using the assessment data could provide valuable information. Below are five potential research questions:

- What is the right technology for a given goal or learner?
- How do learner characteristics, such as media proficiency, mediate the effectiveness of Internet teaching technologies?
- What is the right mixture of technology and social interactions, whether face-to-face or virtual to benefit individuals and society as a whole?
- How can we anticipate the effect of new technologies?
- What are novel designs for technologies that promote learning?

Internet technologies do not just offer an alternative to classroom learning, but a potential to substantially change education. Only with additional research on technology assessment and learner preferences might the idea of Internet-driven education become a significant part of the educational system.

IDR Team Summary 3

Define the trajectory, value, and risk of Extreme Lifelogging when nearly everything about a person is in Cyberspace.

CHALLENGE SUMMARY

As data is ingested from and about our daily lives using smartphones, computers, and other sources, society is on a clear trajectory to have various degrees of "Full Life Recording." "Extreme Lifelogging" is a on a decade course to allow an individual to record everything they see (Caprani et al., 2001), hear and much more with the aid of GPS tracking and on body health monitoring aids (Topol, 2012).

A log of our information interactions is a step toward management and integration of our information and ourselves. A log that includes GPS encoding, near-field interactions with devices, pointers to pictures we've taken, or videos we've recorded begins to approach the fidelity of a lifelog with a potential to serve as a lifelong complement to our internal memory and our digital selves. This lifetime log is the basis for the creation of understanding and stories.

Being able to reliably store *and retrieve* a lifetime of information within personal devices and the cloud is practical, inexpensive, and inevitable. This inevitably raises issues in every aspect from recording though the lifetime use and hereafter life storage that research and dozens of companies are forming to solve.

As we struggle to find, organize, and use life records, one challenge is *information fragmentation* of our personal information or our *cyberself*. We need to know where our cyberself is located, who owns and can access it, and when (now and in the future). Of course control involving permanency, privacy, and security of our digital selves is always a concern as we continue to allow or offer more public access to our cyberself. For people's lives that are maintained by public institutions such as universities and national libraries (British Library, 2009), maintaining life records of hybrid analog-digital people is a challenge. For example, who will be in public digital lifeboats?

We also want our logs to have existence and persistence independent of the applications and devices we happen to be using at any time and held anywhere. **Our digital life is forever.**

Key Questions

Will "extreme lifelogging" actually occur as determined by whether such systems can be built that serve a useful purpose i.e., are able to create a market? What privacy laws or concerns will inhibit their existence?

How can meaningful structures emerge as an effortless by-product of our interactions (with our information and with other people)? E.g., E-mail threads tell useful stories of discussions which extend across and "over" time. "Stories" are perhaps the most useful structures about a person. How can these be constructed automatically?

How much can a person's eMemory help that person in the event of declining function including various neurological diseases?

What are the costs to store my cyberself forever? Assuming I pay for this upfront, how can I guarantee my forever existence without a physical self?

What mechanisms e.g., standards, laws, technology is required to insure the long term accessibility of digital lives such that these personal bits will be always readable?

Will the BCI (Brain Computer Interface) play into such systems? Ideally, a person's eMemory is a person's real and lasting memory, and a URL and metadata to access this content.

Reading

Bell G. and Gemmell J. Total recall: Your life uploaded. Dutton, Penguin: New York 2009.

British Library. First digital lives research conference: personal digital archives for the 21st century, 9 February 2009.

Caprani N, O'Connor N, and Gurrin C. Motivating lifelogging practices through shared family reminiscence. In: CHI 2011 workshop: Bridging practices, theories, and technologies to support reminiscence, 8 May 2011: Vancouver, Canada.

Jones W. (XooML: XML in support of many tools working on a single organization of personal information. In Proceedings of the 2011 iConference 2011: Seattle, Washington;478-488.

Ringel M, Cutrell E, Dumais S, Horvitz E. Milestones in time: The value of landmarks in retrieving information from personal stores. Pr*oceedings of Interact 2003: Ninth International Conference on Human-Computer Interaction* September 2003: Zürich, Switzerland.

Topol E. The creative destruction of medicine: How the digital revolution will create better health care. Basic Books: New York, 2012.

IDR TEAM MEMBERS

- Carole R. Beal, University of Arizona
- Robert M. Bilder, UCLA
- Kim T. Blackwell, George Mason University
- Chris Forsythe, Sandia National Laboratories
- Robert A. Greenes, Arizona State University
- Cathal Gurrin, Dublin City University
- Ning Lu, North Carolina State University
- Ashley Taylor, New York University
- Anthony D. Wagner, Stanford University

IDR TEAM 3

Ashley Taylor, Science Writing Scholar
New York University

IDR Team 3 was asked to define the trajectory, value, and risk of extreme lifelogging.

What Is Extreme Lifelogging?

Lifelogging is simply recording any kind of information about one's life. Older forms of lifelogs include photo albums and diaries. More modern types of lifelogs include Facebook timelines and Twitter feeds. In all these examples of lifelogging, an individual selects moments to record. Extreme lifelogging, as defined by Gordon Bell, a Microsoft researcher, lifelogger, and co-author of a book on the subject, is when you capture "everything you see and hear." Bell thinks extreme lifelogging would be possible by 2020. By Bell's definition, extreme lifelogging is a thing of the future. However, several devices are bringing people closer to extreme lifelogging, and these devices collect information about one's life in a different way than the kinds of lifelogs—Facebook, diaries—mentioned before. These lifelogging devices to collect data about one's life and do so automatically, at periodic intervals or in response to certain stimuli, rather than handpicking moments to record. Moreover, they go beyond Bell's definition by including other dimensions of data about the individual beyond experiences seen and heard, such as movement, location, body function, and interactions with other people or the environment. Two types of lifelogging devices are lifelogging cameras, which are worn around the neck and take photos automatically, and high-tech pedometers that automatically sync to phones and USB drives.

Lifelogging cameras

Two prominent lifelogging cameras are Microsoft's SenseCam and the forthcoming Memoto, made by a Swedish company. The SenseCam is a camera about the size of a deck of cards that hangs around the neck and automatically takes photos depending on how you set it. It was invented in 2003 by Lyndsay Williams, of Microsoft Research. One can program the SenseCam to periodically take photos, for example, to take a photo every 30 seconds. One can also set it to take photos in response to certain environmental stimuli, such as body heat and changes in light levels.

The SenseCam and software to view the photos are currently available only to researchers, such as Gordon Bell, who has worn the SenseCam since 2003 and Cathal Gurrin of Dublin City University, who has worn a SenseCam since 2006 and has logged over 9 million photos. A commercial version of the SenseCam is available as the Vicon Revue; however, it is being discontinued at the end of 2012, according to the product website.

Though some of the hardware for extreme lifelogging is already on the market, the software for extracting useful information from lifelog data is limited. One can buy software to view photos taken by the Vicon Revue. Viewing photos in order creates a flipbook effect. However, the software to tag those images and search them remains in the research stages.

Memoto, founded in 2012 and funded, in part, by a $550,000 Kickstarter campaign, is taking pre-orders for its version of a lifelogging camera and expects to deliver its first batch of orders in April 2013. The camera, a 1.4-inch square box that clips onto clothing, will take a photo every 30 seconds and record where and when the photo was taken. Memoto has iPhone and Android apps for organizing the lifelogging data, and these apps can connect to social networking websites if one chooses. Memoto will store customers' lifelogging data in the cloud for a monthly fee.

Pedometers/physical monitors

Another kind of lifelogging device is a pedometer that records steps taken and other measures and wirelessly uploads them to smartphones or USB drives. One of these pedometers is FitBit One, a device about the size of a flash drive that uses its three-axis accelerometer and altimeter to record steps taken, miles traveled, and stairs climbed, and also to monitor sleep quality. It calculates calories burned. Unlike older pedometers, it syncs automatically to smartphones and to a USB drive and comes with applications for analyzing the data.

The BodyMedia armband is another automatic fitness tracker with more sensors than the FitBit. It records calories consumed, steps taken, and sleep quality, and includes an online Activity Manager and smartphone app. Unlike the FitBit, it also measures skin temperature, the heat dissipated from the skin, and the skin's electrical conductance (a function of sweating) in order to better calculate calories burned, claiming over 90 percent accuracy. With a strap purchased from another company, the device can also measure heart rate.

State of the science

As the above examples indicate, some of the hardware for extreme lifelogging is already on the market, and more is coming. However, the software for extracting useful information from lifelog data is limited. The software to tag and search images recorded by lifelogging cameras remains in the research stages. Generally, tools for aggregating one's own data over time and from different sensing devices are in the development stage, as are tools for aggregating the data of many people. Furthermore, there are no laws governing how lifelogging data will be used and safeguarding, or at least regulating, the privacy of personal data.

To address these deficits, IDR Team 3 came up with several proposals for how to manage lifelogging data, on a personal level, through apps, and on a societal level, by developing regulations to safeguard privacy of lifeloggers and those they capture on camera. Before delving into discussion of how to deal with lifelogging data, the IDR Team began by considering the pros and cons of the practice.

Is Lifelogging a Good Idea?

The IDR Team acknowledged that with so much data collected about people already, through Google, smartphone usage and GPS tracker logs, and our various transactions such as online purchases, for example, life logs about people are already being developed, like it or not, though individuals may not have access to the information. The question is whether people will choose to collect minutely detailed data about themselves, and if they do, how they will use the information

Potential downsides of extreme lifelogging

A digital life log could give a false sense of completeness. Though a lifelogging camera will record one's external experiences in great detail, it does not capture one's emotional reactions. In order to make lifelog photos searchable, they need to be tagged with words, and the tags, in standardizing and categorizing things people see, will fail to capture all the richness of life experiences.

Furthermore, losing some information could actually be a desirable feature of our biological memories, some IDR Team members thought. Forgetting embarrassing moments or unpleasant experiences is a coping mechanism. Do we want a record of everything? Do we want to actively participate in creating a record of our pasts that might come to haunt us?

The IDR Team identified several risks of collecting terabytes of personal data, primarily with respect to threats to privacy. Lifelogging would probably increase the risk and magnitude of identity theft; someone who stole one's lifelogging data would have much more information than the thief could get from someone who was not lifelogging. Employers or insurers, if they accessed the data, might use it to discriminate against people. Would employers ask to look at life log data or provide strong incentives for employees to share the information? Would insurers give discounts to people who collect life log data and share the information with them? Lastly, life log data would infringe on the privacy of people who are not lifeloggers but are recorded in other people's logs, for example, as figures captured by other people's SenseCams.

Potential advantages of extreme lifelogging

That said, the IDR Team saw several potential advantages of lifelogging and identified three categories of stakeholders who would benefit from the practice in different ways: individuals, government/society, and corporations. At the individual and societal levels, the team focused mostly on how lifelogging might be useful to improve health.

Lifelogging could help individuals better understand and predict the consequences of their actions. Whether the information was about eating habits, blood pressure or anxiety, it could help people interested in self-improvement by providing a more accurate picture of how they are now. Our group focused on the potential of extreme lifelogging to improve health, both of individuals and of society overall. The general idea is that if a rational person knew, to take a made-up example, that a certain level of exercise reduced one's chances of a heart attack by a certain amount, that person, using a FitBit, might exercise at the recommended level and improve his/her cardiac health. Individuals improving their health would lower healthcare costs overall, which would be good for society.

That example is, of course, oversimplified. There are several caveats to the idea that extreme lifelogging could help people improve their health. First, how strong are the correlations between behaviors and outcomes? To take the previous example, how certain are scientists that exercise can avert heart attacks? The group did not discuss these kinds of correlations. Theoretically, aggregation of extreme lifelogging data about exercise combined with records of heart attacks could produce evidence in support of such a correlation, and the more data went into it, the more specific the correlation would be. Yet even if aggregate lifelogging data suggested that a person who exercised regularly had a lower chance of a heart attack, that wouldn't change the fact that anyone could have a heart attack at any time. Lifelogging data might tell an individual the most likely outcomes of their behavior, but even then, what is most likely is not always what happens. Furthermore, people are not rational and don't always act rationally on the information they already have about their behavior and health. More data, from extreme lifelogging, may thus not make people more rational. People's tendencies to be irrational would limit the potential usefulness of lifelogging data.

In addition to health monitoring, individuals could use lifelogging as a memory aid, something that would be particularly useful to people with memory loss. Microsoft's SenseCam has been shown to improve the memory of a patient with amnesia, who was better able to remember events after recording them with the SenseCam and reviewing the footage.

Corporations could use lifelogging data to see how people are using their products and generally to observe consumer behavior. People wearing a SenseCam would be collecting data not only about their own lives but also about all the people and things they encounter in their environment. Everyone will collect information irrelevant to them that could be important to someone else. The IDR Team discussed the likely possibility that people could sell their data to corporations.

The IDR Team proposes three ideas for making lifelogging data more and for establishing standards for how the data can be used: a conference, at which people would discuss a Consumer Bill of Rights for lifeloggers; an open-source platform where people could share software for analyzing life log data; and a competition to create lifelogging apps. These proposals, outlined below, would also serve to develop a community of people interested in lifelogging.

Solutions

The IDR Team decided that it would be important to develop a policy framework for laws to protect people from abuses of life log data. For example, the IDR Team thought it would be important to develop laws to prohibit employers and insurers from discriminating against people based on lifelog data, analogous to a 2008 law prohibiting them from discriminating based on genetic data, the Genetic Information Nondiscrimination Act. In order to develop these regulations, the IDR Team proposed a conference where participants would develop a Consumer Bill of Rights for lifeloggers.

In order to allow people to explore correlations in lifelog data and draw strong conclusions, it will be necessary to gather data from many people and aggregate it. The IDR Team proposed developing an open-source platform where people who developed software related to presenting and analyzing lifelog data could share their work. People could post software components that would be useful in making apps for analyzing lifelog data. An open-source platform would aid development of software by facilitating cooperation. It could save individual app developers from creating software in parallel when they could just use existing software and focus their time on developing original programs. In addition to promoting the creation of apps, such a platform would help create a community of people interested in lifelogging.

The IDR Team then proposed a competition to encourage the development of apps that would make lifelog data useful—for example, a fitness app that would use data from physiological sensors to make health recommendations. The IDR Team would provide "contestants" with data, and they would need to develop an app to analyze it.

Future Work

All three proposals are first steps. After development of a Consumer Bill of Rights would come implementation. After the creation of open-source software would come app creation. Once apps were available to the public we could expect the more exciting part: where people start aggregating and analyzing data from large numbers of participants to learn things about human health and behavior. Finally, time will tell if people will act on the information gathered from lifelogging. An outstanding question is what happens to the data when people die, something that's an issue today and will be a larger one as people leave larger digital traces.

Lifelogging of data is a technology that will increase people's information about their lives, and the recommendations of the IDR Team will help people to use that information wisely.

IDR Team Summary 4

Identify the ways in which the Internet positively and negatively impacts social behavior.

CHALLENGE SUMMARY

    The Internet is profoundly affecting how people form relationships, organize, collaborate, help one another, consume and produce information, and make collective decisions. Though forming social networks is not new, what is new is that electronic networks can help people to maintain networks, but also to expand the reach of their networks. The Internet enables new forms of organizing at an unprecedented scale, from creating distributed social groups to mobilizing political action. Online collaborative peer production has created Wikipedia, an encyclopedia that experts have judged to be no different in quality from the traditional print version (Giles, 2005). Citizens have become news correspondents and editorialists, using platforms such as microblogs, blogs, and social-networking sites to report and comment on current events, in many cases faster than traditional news media. When crises occur, people no longer need to depend on formal official responders for aid; people are using an array of Internet applications to locate lost victims, send resources, and broadcast situational awareness about the crisis. Last, through crowdsourcing, people combine small contributions to achieve large effects such as solving complex scientific problems.

    The Internet allows people to be social beings for nearly all their waking hours. The ubiquity of the Internet and social media raises numerous questions about its effects on social behavior across many spheres of daily living. For instance, many young people now bring laptops to classes and iPhones to the dinner table, where they are able to stay connected to their social worlds, through e-mail or social networking. Unknown is how this has affected their ability to attend to the demands of their current situations, such as paying attention to lectures, engaging in conversation with friends and family, and more worrisome, driving. How easily can people disconnect from the Internet? Devices such as mobile Wi-Fi hotspots allow for constant Internet connections, which also leads users to expect near instantaneous responses from others who they assume will be connected. It is unknown how such expectations shape social dynamics and how they may interfere with self-regulatory ability (such as being able to delay gratification).

While the Internet clearly has the potential to enable social collaboration on an unprecedented scale, there are also concerns about how its increasingly central role in social interactions may fundamentally change human society. For example, the Internet provides individuals with the ability to interact very widely with other individuals that agree with them on political issues, which may intensify political polarization and reduce the ability to compromise. We also do not know whether the mental and physical health benefits of social interaction extend to the virtual world. How will concerns about privacy, identity, and deception on the Internet affect how people interact? New research approaches are needed to better understand how human social function is being impacted by increasing immersion in a virtual social world, particularly on the development of social function in children.

Key Questions

- How has technology changed social relationships? What aspects of social networking are good and which are bad?
- Does the availability of online social networking increase or decrease the openness to new ideas? Does online social interaction encourage "assortative friendship" in which individuals interact only with others who agree with them on fundamental issues?
- Does the Internet bring people together or pull them apart? Does the blogosphere lead to greater partisanship and narrow thinking or does it unite the global community and expose people and societies to new ideas? What implications does this have for political systems within nations and for relations between nations? Is the current political climate more partisan and polarized because of the Internet?
- How does connectivity affect the creative process, and how we learn, communicate, process information, and behave with each other face-to-face? Is there a difference between digital natives and immigrants?

- Have social networks, such as Facebook, changed the meaning of what it means to be "friends?" Early theories suggested that online communications would displace and reduce connections to friends and family (see related story in http://chronicle.com/article/Faux-Friendship/49308/). Alternatively, some recent theories have suggested that online communications stimulate and enhance the closeness of relationships, perhaps by leading people to disclose more personal information online.

- How do electronic social networks influence health behaviors, both positive and negative? Social support is an important component of many health interventions and social networks may be implicated in health issues (e.g., obesity).

- How, if at all, does continual connectivity affect skills in offline social interaction? While many studies have addressed online behavior, especially in young people (e.g., Turkle, 2011), few studies have carefully examined the relationship of Internet use and social skills in real life.

Reading

Benkler Y. The wealth of networks: How social production transforms markets and freedom. Yale University Press: New Haven, CT, 2006.

Ellison NB, Steinfield C, and Lampe C. The benefits of Facebook friends: Social capital and college students' use of online social network sites. Journal of Computer-Mediated Communication 2007;12:1143–1168.

Giles J. Internet encyclopedias go head to head. *Nature* 15 December 2005;438:900-901.

Turkle S. Alone together: Why we expect more from technology and less from each other. Basic Books: New York, 2011.

Because of the popularity of this topic, two groups explored this subject. Please be sure to review each write-up, which immediately follow this one.

IDR TEAM MEMBERS—GROUP A

- Alison Bruzek, MIT
- David S. Hachen, University of Notre Dame
- Kenneth M. Langa, University of Michigan
- Kalev H. Leetaru, University of Illinois
- Jeffrey Liew, Texas A&M University
- Aaron D. Striegel, University of Notre Dame
- Yalda T. Uhls, University of California, Los Angeles

IDR TEAM SUMMARY—GROUP 4A

Alison Bruzek, NAKFI Science Writing Scholar
Massachusetts Institute of Technology

IDR Team 4A was asked to identify the ways in which the Internet positively and negatively affects social behavior. Historically, new technology has always changed the way people behave with each other, whether the shift was away from oral traditions to the creation of novels or the invention of television. The team chose to examine this question through the lens of human development in the digital age. Specifically, the group examined the Internet through use of social media like Facebook, Twitter, and YouTube in the United States.

The team listed several unique factors that characterize contemporary social media including: anonymity, speed, instant gratification, self-selecting isolation of viewpoints ("cyber-ghettoization"), 24/7 access, decentralization, large user base, and lack of regulation by authority. One of the major questions posed by social media is whether it is truly bringing us closer together as is so often stated.

How We Build Relationships

It has not yet been determined whether the ways in which people communicate with each other online affects the extent and qualities of people's offline communication. Online activity could affect not just the number of offline social contacts people have but also the content of offline conversations, the emotional character of those conversations, and the exit strategy for any given interaction with an online "friend." In particular, IDR Team 4A looked at the difference between strong ties and weak ties in the offline and online worlds. Weak ties were defined as acquaintances or contacts that could not be relied on as a strong support system, and were less frequent, intense, pervasive, and reciprocal in nature than strong ties.

The speed and scale of the Internet has allowed for the rapid growth of weak ties. There are indications that the Dunbar number (proposed by British anthropologist Robin Dunbar), a count of the average person's social network size, have grown from 150 to 300 or more in recent years. This is also consistent with the idea of Albert-László Barabási's characterization of the Internet as a "scale free" network. While there are several hypotheses for what this increasing number of weak ties means for individuals, IDR

Team 4A focused on the idea that the creation of so many weak ties may perhaps be coming at the cost of developing strong ties, a proposition supported in literature by displacement theory.

Why strong ties are important

Strong ties were defined as relationships that are reciprocal, long-lasting, and characterized by high intensity. Research indicates how important strong ties are for individual mental and physical well-being, as well as for larger societal benefits they bring, including a population capable of empathy. Strong ties build trust and reliability between two people. They allow for the practice of social interactions and because they occur in real-time, can permit expansion of initial ideas and self-reflection. This leads to perspective taking, a critical part of developing empathy. While often based on the individuals' similarities, these real-time relationships offer a strong feedback loop (a way for both participants to receive feedback on their behaviors), as well as an exchange of emotion in a shared physical space.

IDR Team 4A identified several questions that are crucial for the development of strong ties and asked if these traits had analogues online:

- How do we build trust and reciprocity online?
- How do we create online relationships with persistence and reliability?
- How do we "slow down" online time so that people have the ability to self-reflect?
- How can we mimic physical shared space online and the synchronic nature of offline conversations?

Strong ties in a digital world

While acknowledging that the beginnings of strong ties can be created in an online world, Team IDR 4A agreed that in order to fulfill some of the components of strong ties, the online relationship would need to move offline. Complicating this, the team realized that we currently conceptualize offline strong ties as important for personal and societal development, but we do not know to what degree face-to-face communication will be valued in the 21st/22nd century world. Rejecting the idea of technological determinism, it is possible that there is a feedback loop such that the technologies being created for Internet communication merely reflect the ways in which people now choose to communicate. It could be that these changes in people mean that strong ties are no longer the "gold standard" and are being devalued in favor of many weak ties (i.e., YouTube instant fame).

Examining Strong Ties

To investigate the potential displacement of strong ties in a digital world, the following research questions were developed:

- What factors are critical to creating, maintaining, and preserving strong ties offline that can be brought online?
- Are offline strong ties really weakening and disappearing?
- Is the Internet driving/amplifying the demise of strong ties, or is it reflecting broader societal changes?
- Are there positive aspects of strong ties that already exist in some form in the digital world and are being encouraged?
- How can we use the Internet's tools to preserve and maintain strong ties?

Team IDR 4A further examined the first question, specifically how strong ties are developed and how the Internet can facilitate positive behaviors for the creation of these ties. It was determined that one of the several critical factors for strong ties online is synchronous emotional connection, such that the specific level and type of emotion a person exhibits in a conversation is correctly perceived by the other person. Given this emotional component, several other questions arose including:

- How do the mediums, online or offline, meet an individual's needs and goals for expressing emotion—emotion being a way that builds strong ties?
- Does the Internet limit or facilitate the ability to exchange emotion based on individual differences (autism, extroversion, introversion) and how?
- How does gender/age/culture/SES affect the exchange of emotion online?
- Is the ability of children to understand the emotional expressions of others underdeveloped due to increased use of the Internet?
- Does increased Internet use ultimately retard the ability of 21st century children to develop their "emotional scale" in the offline world?

All of these issues could affect an individual's ability to create lasting strong ties both online and offline. From these ideas, two research directions were proposed. The first examines individuals' current state of what IDR Team 4A referred to as "Emotional (Intelligence) Quota" (EQ), a measurable standard for the capacity of people to express and interpret emotion. The second includes potential experiments to identify factors that could help people online grasp each others' emotions.

Studies and interventions

IDR Team 4A recommended more comparative studies on high media users and low media users, particularly analyses broken down by age, socioeconomic status, ethnicity, culture, and personality differences. These studies should examine the differences in an individual's emotional health and the number and strength of their strong ties. For example, there is recent literature to support ideas like "Rich Get Richer" and "Social Compensation." In the former, the online social sphere is beneficial for extroverts who are able to continue their many and frequent offline interactions in the digital world and can even expand the size of their networks and become even more popular online. In the latter, introverts can use the digital world to create new social interactions and the beginnings of new strong ties that would not normally occur in offline life. In addition, longitudinal studies should be done to understand what happens to their networks (strong and weak ties) and mental health or EQ as people start to increase or decrease their online usage.

Based on current knowledge of strong ties and online communication, IDR Team 4A proposed the development of tools to facilitate emotional exchange and identify/promote methods for synchronous communication. This could include items like an "Emotion Scale." This scale may offer insights into how people receive emotions as opposed to how they are broadcast and could give a quantifiable way for people to examine what their EQ is. Simply by knowing if they are deficient, they can perhaps resolve the issue. There is also the technology of sensors or other biofeedback devices to let individuals know each others' specific emotions at specific points (i.e., visual displays). For example, much like a video game might indicate how a particular character is feeling in terms of health or happiness, a similar device might indicate to the world how a person is feeling. Alternatively, a less desirable and ultimately less achievable goal is to decrease online usage, particularly in young children, in order to allow for the development of traits related to expression and perceiving emotion that scientists currently believe are difficult to acquire online.

Future Considerations

Because the Internet is still new, many of the studies considered and proposed have not yet been undertaken. It is difficult to know now whether the relatively new digital world will or will not reflect the offline world's organic creation of strong ties. We do not yet know if the Internet or social media is bad for your health. We cannot say that it is a zero-sum game and that online weak ties are replacing offline strong ties.

However, because individuals differ, it would be most beneficial if we could predict whether the Internet will limit or facilitate strong ties for specific types of individuals and then tailor their use of social media to best adapt to their needs. If we can address the potential issue of changing strong ties as a result of social media, we can improve the emotional, mental, and social health of society and its individuals who are developing in an era of high media consumption.

IDR TEAM MEMBERS—GROUP B

- David C. Cook, Government of Western Australia
- Nicole B. Ellison, Michigan State University
- Eva K. Lee, George Institute of Technology
- Anthony C. Olcott, Central Intelligence Agency
- Eliot R. Smith, Indiana University
- Diane M. Sullenberger, National Academy of Sciences
- Kelly Tucker, Texas A&M University

IDR TEAM SUMMARY—GROUP 4B

Kelly Tucker, NAKFI Science Writing Scholar
Texas A&M University

IDR Team 4B was tasked with identifying the ways in which the Internet positively and negatively impacts social behavior. The rapid evolution of the Internet, online social media, and various networking platforms has led to changes in how people communicate and what norms and standards regulate communication. While communication has changed over the course of human history, the pace of that change has accelerated so much in recent years that individuals less than five years apart in age have very different perceptions of what constitutes a "friend" or a conversation. Determining the impact of the Internet on social behavior is essential to understanding trends and new forms of communication.

Redefining the Question

The IDR Team first determined that the assigned challenge was too broad to tackle in its entirety and that it would be more useful to focus on an aspect of the challenge instead. However, the IDR Team struggled to determine what the focus and scope of such an aspect should be. IDR Team members each suggested topics related to their backgrounds. Potential subjects included the exponential increase in information available to Internet users, the impact of online reviews on consumer choices, the creation of online groups and clusters of like-minded users, and the Internet's role in social and political movements.

Based on the discussion of possible topics, the IDR Team ultimately agreed to examine how users filter what information they take in from and what information they submit to the Internet. In this context, "information" includes data, relationship-related exchanges, and viewpoints. This aspect of the broader challenge also covers topics such as the formation of polarized groups that exist in online "echo chambers" and how users feel that they have a right to have their voices heard online regardless of credentials or background.

The Current Online Landscape

One of the most important features of the Internet is the lack of barriers to information production and distribution. Historically, only a small number of people in a society were capable of creating, distributing, receiving, and processing data and information. The Internet has expanded the availability of these abilities from an elite minority to the vast majority of people. This change has lowered the barriers to information production and distribution. In addition, many individuals now have the sense that each person has a right to empowerment and to be heard online. By shifting these aspects of and attitudes toward access, the Internet has allowed more people than ever to find and utilize information and connect with other individuals.

Research Questions

With a working definition of the redefined challenge and the contemporary backdrop in mind, the IDR Team came up with three principal research questions aimed at filling in knowledge gaps related to the challenge.

What are the effects of these changes?

One of the major negative aspects of the increased flow of information is the large number of voices jockeying to be heard. The credibility of these voices is often unknown and suspect at best. This creates a noisy online environment that complicates decision making. Conversely, these changes have created a virtual space that allows individuals to express their views and be heard by others that would not otherwise be able to listen. Offline research suggests that this may be beneficial by allowing people to better identify with institutions that provide such a platform and may increase the likelihood that people will accept institutional decisions that run counter to their interests.

Potential studies related to this question could take several different avenues. One possibility would be to examine where the threshold for triggering these positive feelings lies. Another study could explore whether these feelings can be caused by merely "posting" or whether there must be some form of positive feedback or encouragement from other users to create positive feelings. A contemporary example of this sort of activity is the rise of online petitions or "slacktivism." Do people who simply electronically sign a petition get positive feelings equitable to, greater than, or less than those of people who physically sign a petition? Do the feelings come from the perception of "doing good" or "looking good?" Does the petition need to be successful in some way or is signing it sufficient? Might these types of easy online activities cause people to feel "I've done my share," substituting for more traditional (and possibly more meaningful) forms of activism such as volunteering for organizations, doorbell ringing, or contributing money?

How do people conceptualize online information filters?

This question highlights the existence of filters that sift through the information that reaches an individual. Information does not have to flow through these filters linearly or in any specific order. The IDR Team identified five major filters:

- Content Creation—the origin/originator of the information
- Social System—online reviewing systems for products and content, tools that monitor the "most viewed" content and designate it as such
- Technical—behind-the-scenes programs that promote content based on user history such as Facebook's algorithm for individual newsfeeds and Google's PageRank
- Source—the individual's choices about what information to see and further process
- Psychological/Cognitive—cultural background and demographics, individual traits that affect what a given user is more likely to look at, and how an individual determines information or a source's credibility and value

Research exploring this aspect of the Internet's effect on social behavior would first have to ascertain to what extent people are currently aware of these filters. For example, are people aware of how Google personalizes each user's searches and how Facebook determines what appears is in an individual's newsfeed? After determining this baseline, further studies could explore the strategies users implement to manage filters and whether increased filter awareness could alter their behavior in seeking, evaluating, and using information from the Internet. If behavior can be altered, what sort of intervention would be both appropriate and effective?

What are the implications of these filters?

The existence of filters seems to indicate that they may be responsible for the creation of online "echo chambers." These chambers are formed by users that seek only those individuals or voices that have similar interests. This phenomenon deters online debate and access to new information and encourages like-minded individuals to cluster and feed each other's existing views. The resultant positive feedback loop creates the "echo chamber" where likeminded users only hear what they already know or what they are likely to agree with. Users that never leave the "chamber" will not have the opportunity to encounter different or opposing viewpoints and may never encounter material or ideas that could spark a new interest or idea.

On the positive side, filters may be able to amplify existing relationships between individuals with different viewpoints. For example, if a Facebook user's friend posts an article and the user comments on it, the poster's other friends can see the user's response and respond to it. By virtue of having a mutual friend (the poster), the user and poster's friends can interact and debate. If such is the case, filters would actually help moderate online discourse and/or increase tolerance in the context of friendships or other relationships.

The Benefits of Understanding Social Behavior and the Internet

The IDR Team concluded their discussion by examining the broad benefits that may result from a better understanding of how the Internet affects social behavior. The group identified governance (not to be restricted to government), education, and civic participation as the three major areas that might benefit from the aforementioned research. The IDR Team expects that even more research and study questions would emerge from those posed here. Such questions might include:

- Where is government authority derived from in this changing environment? How do we regulate social media systems? Who is the agent of change?
- How is expertise determined and signaled? Who has the authority to design and implement curricula?
- Whose voice counts? How are voices counted?

IDR Team Summary 5
Develop a new approach to assess the differences in cognitive and brain function between the brains of digital natives and digital immigrants.

CHALLENGE SUMMARY

The world in which most humans live today is radically different from the one in which the human brain evolved. Technology has enabled an informational environment that exposes individuals to an amount of novel information each day that is orders of magnitude greater than the amount of novel information experienced by the ancestral human. In this challenge, participants will attempt to understand how the mind and brain adapt to the modern informational environment.

The information processing limitations of the human brain are well known; in particular, there appears to be a bottleneck in the decision making process that limits the ability to truly perform multiple cognitive tasks at once. Further, it is now clear that there are serious public health consequences (such as increased automotive accident rates) associated with attempts to multitask. Research has begun to address the consequences of widespread use of electronic devices (known as "media multitasking"), but we do not currently understand its implications (either positive or negative) for important brain functions such as learning, decision making, motivation, and emotion. There is also particular concern regarding the effects of informationally-driven cognition on reflection, contemplativeness, and conceptual thinking.

Of particular interest is the question of whether children who develop within an informationally-rich environment (so-called "digital natives") differ in fundamental ways from individuals who only experience these environments later in life ("digital immigrants"). However, this question is extremely difficult to address using controlled experiments, and comparisons between cultures or subgroups that differ in media exposure will necessarily be highly confounded. Despite this difficulty, understanding the effects of informational overload on brain development is critical if we wish to know how the human mind and brain are adapting to the world as it currently exists and whether there are particular approaches that would allow individuals to better adapt to this world.

Key Questions

- How does the brain process the constant barrage of information individuals are exposed to every day? Are the brains of "digital natives" and "digital immigrants" different in the way they process information/expectations, or are these differences cultural?
- What impact does media multitasking have on the ability to synthesize, evaluate, and recall information, especially in stressful situations (e.g. medical emergencies)?
- What kinds of processes/tools can facilitate building of knowledge, conceptual thinking, comptemplativeness, and reflection in a digital age?
- What types of neuroplastic change are occurring in today's "wired" brains that can be capitalized upon to benefit individuals and society?
- Can one improve or retain cognitive and perceptual abilities by mental or physical exercise, and what are the mechanisms by which such improvements are achieved? How can one take advantage of the capacity of the adult brain to undergo experience dependent plastic change?
- How can we create an environment which will pre-dispose the brain to react in ways we consider ideal?

Reading

Junco R and Cotten SR. Perceived academic effects of instant messaging use. Computers & Education 2011;56:370-378.

Ophira E, Nass C, and Wagner AD. Cognitive control in media multitaskers. Proc Nat Acad Sci 24 August 2009; early edition.

Strayer DL, Watson JM, and Drews FA. Cognitive distraction while multitasking in the automobile. In: The psychology of learning and motivation volume 54. Elsevier, Inc. Academic Press: Waltham, MA, 2011.

Because of the popularity of this topic, three groups explored this subject. Please be sure to review each write-up, which immediately follows this one.

IDR TEAM MEMBERS—GROUP A

- Mark S. Cohen, University of California, Los Angeles
- Mark W. Lenox, Texas A&M University
- Andreas Malikopoulos, Oak Ridge National Laboratory
- Rene Marois, Vanderbilt University
- Ulrich Mayr, University of Oregon
- David E. Meyer, University of Michigan
- Jonathan Z. Simon, University of Maryland, College Park
- Clara H. Vaughn, University of Maryland

IDR TEAM SUMMARY—GROUP 5A

Clara Vaughn, NAKFI Science Writing Scholar
University of Maryland

Introduction: Framing the Task

IDR Team 5A was asked to develop a new approach to assess the differences in cognitive and brain function between the brains of digital natives (individuals born during or after the introduction of current digital technologies) and digital immigrants (those individuals born before the widespread use of digital technologies).

The group first developed a framework through which to conceptualize brain function as it relates to the digital world. The framework accounts for potential contributing factors, including the individual under consideration, the technology with which he is interacting and the environment in which he is operating.

The goal of using such a framework was to optimize an individual's cognitive and brain function in the face of an information-dense digital world. It took on the acronym "DITF" (dĭ-tĭf), an abbreviation of OM-EPT DITF, or Optimal Managed Environment-Person-Technology Digital In-Trans-Formation.

Digital "In-Trans-Formation" forms the foundation on which the rest of the concept is based, and was named to characterize the constant flux of digital technologies and information shared across online networks.

In our often transient digital world, the individual must not only deal with a constant flow of information across online channels, but the ebb and flow of new technologies. Optimal management of an individual's cognitive and brain functions as it relates to technology, then, must be highly adaptive to maximize these functions.

To begin developing a practicable management plan for optimizing brain function, the DITF framework considers three key factors: the person, technology, and environment.

The person: Brain-Mind Index

In the framework, the individual is encompassed in a Brain-Mind Index (BMI) that can be used to categorize individuals into types. This sort of personalized cataloguing would allow for individually tailored management that best maximizes the individual's ability to cope in an information-dense digital world.

The BMI would be determined by examining an individual's characteristics including distractibility, working memory, sensation seeking, adaptability and awareness. Other influencing characteristics would be incorporated as the framework is tested and refined.

The technology: Device Complexity Index

One factor directly affecting an individual's cognitive and brain functions is technology, represented in the framework as the Device Complexity Index (DCI). This includes the number of purposes for which the technology is being used, the stability of digital platforms and the sustainable rate of device change, and would incorporate other factors as researchers refine the framework.

The environment: Environmental Confusion Index

The final node in the Environment-Person-Technology triad is the environment in which an individual using technology is immersed. The team named this the Environmental Confusion Index (ECI) to represent the number of potential complicating factors outside of individual-computer interactions that may affect those interactions. The physical, social, cultural, and electronic characteristics of one's surroundings represent factors that might shape an individual's brain-computer interactions.

Optimal management: Adaptive strategies for digital fluency

Developing means to measure environmental, individual, and technological factors and their interplay would be advantageous through providing baseline data that might allow researchers to design personalized strategies for maximizing brain and cognitive function.

Team members proposed management options from digital literacy training starting in elementary school to computerized multitasking management devices as strategies for achieving this goal. Other strategies are discussed later in this paper. They stressed that any plan to enhance brain and cognitive function and efficiency must be adaptive to be effective in helping users navigate the transient digital world.

Concretizing Concepts: Testable Hypotheses

To narrow the focus of the team's examination of the brain and cognitive functions of digital natives and digital immigrants, team members developed *three testable hypotheses* based on the DITF model.

The first proposed that immersion in World 2.0 —the current digital milieu—strengthens processing of salient multi-media cues in sensory cortical sites, leading to stronger bottom-up signals and to disruption of activity in the neural network that supports goal-directed behavior.

This hypothesis follows the line of thought presented by conference speaker Clifford Nass, whose research suggests that high media multitaskers use more brain activity, have greater breadth orientation but lower depth orientation, and tend to avoid deep thinking.

The second hypothesis proposed that chronic multitasking might lead to automation of some brain processes. In other words, high media multitaskers' brains may adapt to the demands placed on them over time, such that these individuals are able to accomplish tasks more efficiently.

The third hypothesis, based on drug addiction research and existing evidence that online "rewards" and drug-induced rewards manifest in similar parts of the brain, posits that chronic exposure to instant multimedia rewards may lead to reduced sensitivity to such rewards, potentially permanently, thereby increasing the individual's reward-seeking behavior through online activities.

Challenges and limitations

All three hypotheses could be tested using functional MRI studies and other brain imaging procedures. Scientists in the group agreed that experiments designed to test the three hypotheses developed during the conference would be fairly easy to carry out, but acknowledged several challenges that would impede progress.

First, measures of outcome and assessment must be developed. *How should efficiency of cognitive and brain function be defined and captured? When developing personalized management strategies, how can personality traits, such as distractibility, be measured?* These and other questions must be answered before meaningful research can progress.

Another challenge is to establish causal, rather than correlational, relationships between behavioral outcomes and cognitive and brain functions. Current research in the field tends to capture only correlational relationships, posing problems for scientists seeking to design brain function-maximizing management plans based on these observations.

Finally, researchers need long-term longitudinal data to establish trends in brain and cognition as technology users adapt to an information-dense digital world. A member of the IDR team said:

> I think that we have a pretty long way to go before submitting a proposal for neuroimaging. The questions that we posed are not of the class, "Where in the brain is X processed?" or "How is multitasking mediated in the brain?" Instead, we ask more longitudinal questions about brain changes with continued behavior. There is an implicit long-term study, or a careful case (of) control design.

Adaptive Strategies: A Model for Maximizing Efficiency

Team members designed questions to approach discussions on digital technologies and brain and cognitive functions, then narrowed their focus to one key question: "Can we develop digital and/or behavioral adaptive strategies to enhance efficiency in World 2.0?"

Such strategies could approach the task of maximizing efficiency in an information-dense digital world either from the human-side (e.g., "technology coaches" that guide individuals in best management practices for digital multitasking) or technology-side (e.g., programs that monitor behavior and apply strategies for improving the technology user's efficiency). Many of these proposed strategies presented inherent application problems. Digital coaches would be costly and most individuals would choose to spend their money elsewhere, for example.

One practical solution proposed by the team was a computerized task monitor that would present visible feedback on time usage (See figures 1 and 2). To use the monitor to maximize efficiency while on the computer, an individual would enter work and personal objectives, and the monitor would track desired outcomes against actual time allocations on the computer. To help goal versus actual time-use distributions align, team members proposed pop-ups that warn users when their time spent on e-mail exceeds goals, for example, or even automatic shut-down of "distraction" windows.

Conclusions

IDR Team 5A refined the lens used to assess the cognitive and brain functions of digital natives and digital immigrants and identified some key challenges in carrying out experiments in the field.

The need for better measures of outcome (e.g., How does the brain perform? What is the comparative efficiency of task-management after adaptive strategies are applied?) and of assessment (e.g., What is efficiency? How is it measured?) were apparent during discussions.

After such measures are established, the need to develop causal relationships between behavioral outcomes and cognitive and brain functions is of primary importance. Only by extracting causality from a series of outcomes can scientists develop adaptive strategies for maximizing brain and cognitive function in World 2.0. The team's goal of maximizing brain function in a transient, information-dense digital world can then be met through cultivating optimizing management strategies, both technology- and individual-driven.

The team presented the several testable hypotheses described above and proposed practical management strategies in its two days of interdisciplinary discourse. Future work stemming from these hypotheses will help scientists better understand the human brain as it relates to the digital world, providing a foundation for developing real-world strategies to better manage the brain in World 2.0.

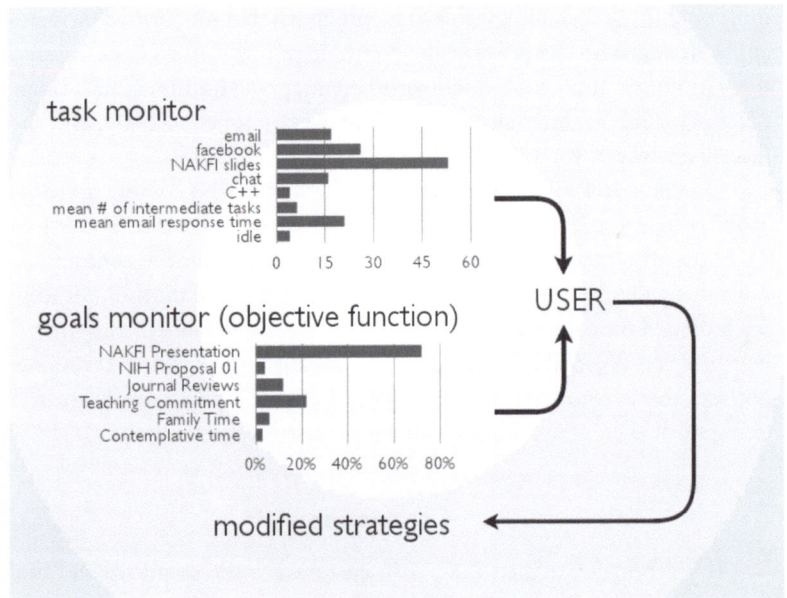

FIGURE 1

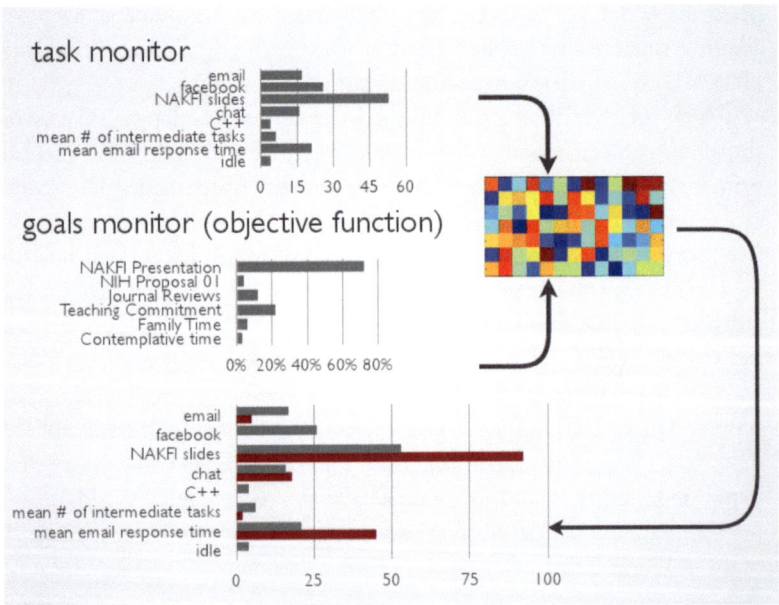

FIGURE 2

IDR TEAM MEMBERS—GROUP B

- Lisa Aziz-Zadeh, University of Southern California
- Steven Kotler, Flow Genome Project
- Jessica Luton, University of Georgia
- Narayan B. Mandayam, Rutgers University
- John Devin McAuley, Michigan State University
- Kimberly F. Raab-Graham, University of Texas at Austin
- David L. Strayer, University of Utah
- Jason M. Watson, University of Utah

IDR TEAM SUMMARY—GROUP 5B

Jessica Luton, NAKFI Science Writing Scholar
University of Georgia

IDR Team 5B was asked to assess the differences in the cognitive and brain functions in the brains of digital natives and digital immigrants. While the initial goal was to address these differences, it became apparent throughout the course of the first group discussions that the illusive nature of defining digital natives and digital immigrants made it difficult to answer this question, as we are all destined to be digital immigrants at some point throughout our lives, when the next big technological advance comes along to which we are not yet accustomed. For baby boomers, the new technology was the Internet, social media, iPads, and iPods. For their parents, that technology was the invention of the television or even the radio. And for today's wired youth, there's no telling what new technology will emerge that they will one day have to learn to adapt just as their parents did. However, the question remains: What exactly is the consumption of digital media, information overload, and constant multitasking doing to the actual chemistry in our brains? How is this changing the way our brain is working? How does our brain work with this technology? And furthermore, are there critical periods of development in which the use of these technologies might have positive or negative effects on the development of a child's brain? IDR Team 5B grappled with each of these questions. However, by the end of conference, only one thing remained apparent: we have no idea how digital media are really affecting the human brain. There's no baseline scientific data that can help answer these critical questions about the impact of technology on brain development. Therefore, a large scale, longitudinal study is necessary.

Media Multitasking and Consequences: A Contemplative Discussion

In thinking about the topic at hand, the team members had a unique set of perspectives that brought about a multidisciplinary discussion of the various consequences that media multitasking and a general consumption of too much media could have on not only social skills, but emotional development and the actual development and inner workings of the human brain. Sociological perspectives, combined with the fields of cognitive neurology and network engineering, led to an enthralling discussion on the topic of media use in the modern world.

In talking about the topic, several assumptions, themes, and questions emerged by the final session. Everyone agreed that we are currently living in an increasingly multitasking world, with a faster pace than was previously thought possible. With that assumption in place, the group agreed that there are inherent limitations to multitasking including personal safety, as is the case when drivers use smartphones in their cars, and the ability to complete a task well when one is constantly interrupted from trying to accomplish the task at hand. The team also agreed that high levels of multitasking may produce impairment in not only completing tasks, but also in development of social and emotional skills. For example, the team discussed the topic of disembodiment. Children develop emotional processing by watching their parents and friends, imitating them and learning to decipher emotions, thereby learning to have empathy for others. The group discussed the implications of what less imitation might mean for emotional development. If children are using text via cell phone and social media, does this mean that they're not developing the skills necessary to understand emotions in real rather than digital people and furthermore experience empathy for other human beings?

Managing Outcomes: The Case of Cognitive Overload

While team participants agreed that very little scientific data actually shows what can be observed in social science research on this topic, the hope is that a longitudinal study might help create solutions for negative and positive outcomes that are observed through this kind of study. That being said, the group discussed several possible solutions to the idea that we are, in fact, experiencing cognitive overload that is impairing our ability to concentrate on one specific task for a long period of time. The use of a queue, whereby secondary tasks are scheduled so as not to interrupt a primary task, is one possible solution. Studying the rhythmic patterns of activity of individuals might allow further insight. If, for instance, tasks are delayed to interrupt at only specific time intervals, say every three minutes, rather than at random times, one might be able to better be more productive. Further solutions might take into account the context of an individual. For example, when a driver is in the car, his or her cell phone could automatically go into a "busy" mode. Those calling would be asked to call back because the driver is busy. In addition, assessing bio or neuro feedback patterns could also help schedule tasks for more productivity. An application that would give the user cues when they're getting off task might also be helpful in getting people back on task.

Beyond those possibilities that are mainly device oriented, one area was discussed as being a possible solution for cognitive overload—training. Because the prefrontal cortex requires more energy to process activities that are not novel, the idea of outsourcing or offloading activities to other parts of the brain via training was suggested. For instance, if one is practiced in specific tasks, those tasks no longer require as much energy for the brain to process and the prefrontal cortex is then left to do other tasks.

For IDR Team 5B, the question remained: How can multitasking performance be optimized without enabling or further contributing to impairments? The team concluded that answering any of these questions is a difficult and almost impossible task without having scientific data from a well-funded, longitudinal study that assessed the differences in brains of high multimedia users and low multimedia users. Using the various disciplines, the group constructed a study format that would allow for this type of assessment. Most important, the team wants to know whether there is a critical period of development in which high multimedia use might inhibit normal cognitive and brain development. The group's proposed study, therefore, focused on various factors including functional differences, shifts in intrinsic oscillation frequencies, early EEG markers and a task battery that would help assess this question from various perspectives.

Constructing a Study: Assessing Cognitive, Social, and Emotional Brain Development

Given the need for a study that would show the actual physiological effects of media use on the brain and therefore help better assess social and emotional effects, IDR Team 5B went about constructing a large-scale cross-sectional longitudinal study that would incorporate a whole range of assessments to help answer some of these difficult questions. While the results obviously wouldn't be available for some time, the implications of initial results could be interpreted as each new cohort advanced in the study. Additionally, the design of such a study, using a new cohort every two years or so, would allow the previous cohort to serve as the control group for the new group. A study of this design, despite being long term and likely expensive, would also allow researchers to study the effects of new technologies as they emerge over the course of the next decade or longer.

In assessing the topic at hand, the team had to come to some sort of consensus on how the terms digital native and digital immigrant might be defined. It didn't take long to figure out that the term is not ideal. Perhaps, the team suggested, media use is on a continuum and a whole range of questions need to assess the type of media being used by different populations, as well as the frequency of media use within the population. Social media, music, Internet, E-mail, talking on the telephone, using a cell phone application, texting, games, sensors like Fitbit, digital TV and video, video chatting, virtual reality, and new emerging technologies yet unveiled were all named as possible categories of media use that might be used as a basis for evaluating media use among today's population.

Beyond looking at the types of media use, the team also sought out a way to assess just how these different types of media are being used or not used by each new cohort. The team constructed a list of analog interactions, such as face-to-face meetings, books and other print materials, and recreational activities that might assess the ratio of digital media use to physical world interactions.

The team also chose not only to look at the types of media, but also the amount of time spent using each media, how many devices are being used simultaneously, what proportion of one's life individuals have been using specific media, when they began their media use, the ratio of digital and face-to-face interactions and the attitudes towards different types of media by study participants.

Defining the (New) Normal

To really provide a broad-based study to help answer some of these questions, the team proposed a hypothetical study that would not only provide a behavioral assessment of study participants, but also investigate physiological measures that might provide a base of knowledge for all disciplines to assess the potential positive and negative effects of high digital media consumption. In this way, the team said it wanted to construct a taxonomy to begin looking at the topic.

The team proposed a study of behavioral characteristics using a whole range of resources including, but not limited to the NASA Task Load Index, an assessment of executive control; testing emotional recognition, processing and management; delayed gratification, timing of interruptions, and implicit imitation; a baseline personality reading; creativity measures, social processing, empathy measures and human vs. machine identification. Beyond these measures, the study would also include a series of physiological measures that might help nail down definitive data on brain development and multimedia use. The team agreed that such a study should look at EEG and MEG data, with specific attention to event related potentials and frequency and time domain. In addition, the study should also use MRI and fMRI data to assess structural brain development and changes, the default resting activity network in the brain, as well as task-related functional connectivity. Furthermore, assessing autonomic responses through ANS, as well as levels of hormones, such as cortisol and adrenaline, might reveal more about the chemistry of the brain when users engage in multimedia use.

IDR TEAM MEMBERS—GROUP C

- David Badre, Brown University
- Ann E. Christiano, University of Florida
- Art Kramer, University of Illinois
- Annie Lang, Indiana University
- Taosheng Liu, Michigan State University
- Oded Nov, Polytechnic Institute of New York University
- Karin A. Remington, Arjuna Solutions
- Rina Shaikh-Lesko, University of California, Santa Cruz

IDR TEAM SUMMARY—GROUP 5C

Rina Shaikh-Lesko, NAKFI Science Writing Scholar
University of California, Santa Cruz

IDR Team 5C was asked to develop a new approach to assess the differences in cognitive brain function between the brains of "digital natives" who have been exposed to the Internet and other digital media since early childhood and the brains of "digital immigrants" who were first exposed to digital technologies later in life. There is evidence that digital natives are more likely to be heavy media users, and more likely to be media multitaskers.

Clifford Nass of Stanford University, in his keynote address, provided evidence that multitasking is a problem that is on the rise among all age groups, but especially teenagers and college-age students, because there are more and more media channels vying for our time. The number of distracting tasks has risen as the number of digital devices has increased. TV, radio, and Internet are now joined by smartphones, tablets, and social media.

There is recent evidence that the vast majority of people—97.5 percent by one measure—are not effective multitaskers. There is also further evidence to suggest that the people who rate themselves as good multitaskers are the worst at it.

However, to IDR Team 5C, the distinction of digital natives and immigrants was an arbitrary one because today's digital natives will become tomorrow's digital immigrants. Heavy digital engagement is simply part of the milieu in which we find ourselves, regardless of whether we are digital natives or immigrants or whether we are heavy or light media users. The challenge, now that we are becoming aware of the potential costs of media overload, is how to manage the competing priorities in an effective way.

Structuring the Environment for Productivity

IDR Team 5C team members think there is a qualitative difference between what the team called "good multitasking" and "bad multitasking." The team defined good multitasking as two or more tasks completed with the same accuracy more quickly and more enjoyably than both (or all) sequentially. An example is when a person downloads audio book chapters onto an iPhone to take with her on a walk to the grocery store. By adding the task of listening to the podcast to walking to the store, and by adding getting exercise by walking instead of driving to the store, three tasks are completed sooner and more enjoyably than any of them alone. "Bad multitasking" is when doing two tasks together takes longer than doing them sequentially and when the process is stressful and unpleasant.

One challenge is that current models of multitasking efficiency rely heavily on cognitive thinking; the dimension of emotion (e.g., enjoyment, pride, stress, fear) is often missing from assessments, as is creative thought. Members of IDR Team 5C speculated there is even a trade-off whereby if the quality of work doesn't suffer, some people are willing to take more time to complete a less enjoyable task if it can be coupled with one that is more enjoyable.

In order to create more opportunities for people to do good multitasking, it is necessary to better understand how the dimensions of pleasure and creativity impact our current understanding of multitasking behavior.

Whistle While You Work

IDR Team 5C proposed taking advantage of technology to track how productivity is affected by the rising number of distractions, digital or otherwise. They sketched basic functions of an app-based study to track productivity and satisfaction with a given app-mediated work schedule. The app is called "Whistle" from the song, "Whistle while you work," reflecting the team's emphasis that the tool provide more than mere time-management tools, but instead, tools that improve happiness and creativity along with productivity.

The app pulls from neuroscience and cognitive science to build an algorithm that provides a schedule for a productive day. For users who download it, the app functions as a life structuring coach, corralling the large number of tasks, distractions, and competing priorities. One team member described it as FitBit for productivity, after the popular, compact smartphone-based health and fitness monitoring tool.

The app would take the information provided by the user and combine tasks that are naturally complementary: listening to classical music while writing versus listening to podcasts while filing. The app can also take into account whether there are times of day that are more productive for certain kinds of tasks. It would be capable of linking a user's calendar, file storage center (i.e., dropbox, hard drives, or cloud storage). The app would also be capable of learning about users with every completed task.

Tapping into the mobile phone app user audience

One of IDR Team 5C members had a recent relevant example of developing, with the aid of a game designer, a free iPhone game app called BrainBaseline. It has more entertaining and engaging versions of cognitive assessment tests used in laboratory settings. The game was subsequently offered on the iTunes store to the general public where within months thousands of users downloaded the game, played it and uploaded their scores, which researchers then were able to analyze.

One of the terms of use will be that the app is free as long as users upload experience data to a central repository. Data uploaded by users can be used by researchers to provide insight about task structure and environments. Researchers can also track how different groups use the app: children, teens, young adults who are new to the workplace, middle-aged workers, managers, older workers, and retirees. Based on information provided by users, researchers can also analyze users by socio-economic class, education and potentially, some health measures. Not only would users be able to view and track their own productivity, researchers would have access to a very large source of participant data on how people work and what it takes for them to be feel productive and happy.

Whistle is essentially a "smart assistant" that provides insights into how an individual user works best and eventually, as data is aggregated, can be applied to larger groups. That larger group data can be accessed by researchers as a rich mine of information on how people function and even thrive in a distraction-rich, resource-poor environment.

Two ways to use The Whistle, two ways to work

There are two distinct ways that the team identified that users could interact with The Whistle app. The first is during the initial set-up—a new user would be presented with a detailed questionnaire which would ask about priorities, tasks, what is enjoyable about those tasks, what are the most tempting distractions and any other relevant information the user wants to add that can all be plugged into an algorithm that would work with a user's calendar, music library, and file storage system to create a daily schedule that is uniquely optimized for that individual.

An alternative approach is for the app to log what people do throughout the day. They could assign a 5-star scale rating on how good or bad they feel about the day in terms of successful task completion and enjoyment. Once there is bank of user data from which to extrapolate, researchers could offer users a four- or five-star day from another user as a way to "try on" new ways of working.

Next Steps

Although researchers have growing evidence about the way the brain reacts to multitasking challenges in an experimental environment that is on a time-scale of seconds to minutes, there is not much evidence of how people react to a rich media environment in a non-experimental everyday setting. We also don't know how the brain functions in a task-rich environment over hours, weeks, or lifetimes.

The Whistle app will begin to address these gaps. As the repository of user data grows, researchers will get a finer and finer grain picture of user priorities and productivity. It will become a catalog of how users want to allocate their time and the way they actually do. Understanding the factors underlying success task management has several potential long-term benefits. Companies could see cost savings if their workers are using a product like The Whistle. There would likely be less stress-related absenteeism and attrition if workers were productively and creatively engaged, which in turn could lead to lowered health insurance costs.

It is likely that the number of digital distraction will continue to grow. Understanding how humans manage multiple tasks is a relatively young field, but it also represents an opportunity to apply our knowledge of computer science, cognitive science, and neuroscience to understanding a problem that is unique to our digital age.

IDR Team Summary 6

Determine how the effects of the digital age will improve health and wellness.

CHALLENGE SUMMARY

As noted in IDR Challenge 1, healthcare systems offer a major test ground for the implementation of methods for efficient lifelong learning that could benefit society through improved health and wellness. The challenge is the cost of doing so effectively and the willingness of care providers and care consumers to adapt their practice and lifestyles to innovative but demanding new technologies. Two potential areas of opportunity are 1) improved career-long education of care providers that would expand the primary care giver roles beyond physicians; and 2) improvements in physician-patient communication for enhancing prospective health strategies. The team challenge is to examine current strategic efforts aimed at these or other comparable health and wellness endpoints and devise practicable means to exploit the digital information explosion in these proposed solutions.

Improved Career-Long Education

In current medical education, the medical student must learn not only the relevant facts and their application to disease mechanisms, treatment, diagnosis and prevention, but assimilate into that body of working knowledge all the new facts that will emerge during their careers as practicing physicians. Most such practice emphasis is devoted to solving acute problems in ill health, injuries, infections, acute cardiovascular or cerebral pathology, or persistent functional problems such as diabetes mellitus, obesity, and epilepsy. This crisis-directed practice may also confront tomorrow's clinicians, but given the growth in biomedical understanding of disease mechanisms and the social, genetic, and environmental factors that can tip the odds from vulnerability to resistance to these disease conditions, how can digital information help? Snyderman and Williams have suggested a strategy that would expand the care team from physicians only to include paraprofessionals (physician assistants, geneticists, epidemiologists, and information specialists) who will be needed to assimilate into practice ever more rapid medical discoveries. How will the digital information explosion be refined into the knowledge needed to enhance the likelihood of success for this strategy? Given the reductions in hours allowed by the Accreditation Council for Graduate Medical Education, the training career opportunities for post-graduate medical education has become seriously constrained by limits in maximum hour work weeks, reducing the time to develop experiential competencies in the skills needed for effective practice, a problem that will be even more critical if the ultimate national healthcare plan reduces the Medicare contributions that presently fund for post-graduate clinical training.

The rapidly broadening armamentarium of powerful new medications requiring lifelong dosing and their complex interactions with individual patients creates multiple potential adverse drug interactions, specific to individual patient diagnoses. While the IBM-Watson project and other expensive, proprietary differential diagnostic systems, are beginning to enter some forms of managed health care, can such computer-assisted diagnostic judgments become an acceptable form of medical practice? Therefore, the underlying problem remains of devising a medical educational system that will not only motivate students to become skilled in basic academics and in the technology of medicine, and remain able to assimilate new knowledge, new medications and new forms of medical practice.

Improved Physician-Patient Communication

If a goal of modern medicine is to improve the general state of societal health, the strategy suggested by Snyderman and Williams calls for enhancing prospective health strategies, by which any individual would maximize their opportunities for lifelong wellness by a team of health practitioners who can implement strategies for disease avoidance based on new knowledge in the genetic, environmental, and social factors that can determine disease onset, progression, or resistance. Levinson and Pizzo have called attention to some of the ways in which current and future patient-physician communication could be improved. Much of such enhancing communication has traditionally been face-to-face in office or bedside, but the onrush of digital information technological options could fragment this budding communication option. If the financial inducements (from the pool of Medicare finances) being offered to hospitals, group practices, and individual practitioners require the demonstration of meaningful use of electronic health records, how can digital technologies avoid becoming a barrier between the doctor who is entering patient details of complaints, findings, and treatment history—while still finding the time to listen to the patient's concerns? How effectively can the Office of the National Coordinator for Health Information enforce that such electronic health records will be interoperable across medical practice systems while ensuring confidentiality of individual personal details and vulnerabilities, and at the same time serve as a national epidemiological surveillance for the emergence of communicable diseases or adverse drug effects?

Improved Health Management by Physicians and Patients

Another major shift in the practice of medicine is the development of digital communication systems to administer medical treatment at a distance, educate patients, and monitor disease states, often termed "telemedicine." An extreme example is the tele-intensive care unit (tele-ICU; Goran 2012). The critical "life-or-death" importance for correct testing, diagnosis, treatment, moment-to-moment monitoring, and constantly reacting to changing health issues places tremendous demands on the multi-modal digital communication system and the human team. How can information, knowledge, and expertise from a high-end hospital improve outcomes for linked ICUs lacking such expertise? In less tense applications, medical management of a disease from a distance is becoming more and more frequent and health-consumer directed. This is due to the availability of accurate patient-worn sensors for blood pressure, heart rhythms, and blood metabolites—not only in daily monitoring but in educating subjects to monitor their blood sugar levels, calculate carbohydrate content of foods they eat, and understand the effects of exercise on insulin utilization, and prevent (or reverse, if needed) hypoglycemic episodes, etc. As in the tele-ICU, it is the integrations of the technical digital components with the human-interaction components that pose the real challenge for the effectiveness and efficiency of system development, especially optimizing the human aspects. How can all the various aspects be integrated to a seamless operation? How can we quickly identify areas of friction (machine/human) and quickly fix them? What would be the properties of an ideal tele-medical application?

Key Questions

- How will increasing technology shape health and medical decision making?
- How will the digital information explosion be refined into medical training?
- Will computer-assisted diagnostic judgments become an acceptable form of medical practice?
- How does medical education motivate students to use digital information to be life-learners of innovations in medicine?
- How can digital technologies avoid becoming a barrier between the doctor and patient?
- How can electronic health records be used to maximize patient outcomes?
- Can information, knowledge, and expertise provided through digital communication systems improve outcomes for those in the field or for those who lack such expertise?
- What would be the properties of an ideal tele-medical application?

Reading

Brailar DJ. Guiding the health information technology agenda. Health Affairs 2010;29:586-595.

Goran SF. Making the move: From bedside to camera-side. Critical Care Nurse 2012;32:20-29.

Halamka JD. Making the most of federal health information technology regulations. Health Affairs 2010;29:596-600.

Levinson W and Pizzo PA. Patient-physician communication. It's about time. JAMA 2011;305:1802-1803.

Moffett TE, Arseneault L, Belsky D, Dickson N, Hancox RJ, Harrington H, Houts R, Poulton R, Roberts BW, Ross S, Sears MR, Thomson WM, Caspi A. 2011. A gradient of childhood self-control predicts health, wealth, and public safety. Proc Natl Acad Sci U S A 15 February 2011;108(7):2693-8.

National Council for Patient Information and Education website.

Snyderman R and Williams RS. Prospective medicine: The next health care transformation. *Acad Med* 2003;78:1079-1084.

Because of the popularity of this topic, three groups explored this subject. Please be sure to review each write-up, which immediately follow this one.

IDR TEAM MEMBERS—GROUP A

- Robert J. Davenport, Brown University
- Margaret Y. Mahan, University of Minnesota
- Todd J. McCallum, Case Western Reserve University
- Paromita Pain, University of Southern California
- Parthasarathy Ranganathan, Hewlett Packard
- Sam R. Sharar, University of Washington
- Tian Zhang, Duke University Hospital

IDR TEAM SUMMARY—GROUP 6A

Paromita Pain, NAKFI Science Writing Scholar
University of Southern California, Los Angeles

IDR Team 6A was asked to determine how new tools and metrics of the digital age will improve health and wellness. It was a very diverse team consisting of a technologist who works with acquisition and interpretation of "big data," two medical doctors and other experts in gerontology, hematology, brain science, and computational biology. The approach from the start was to focus attention on the person at the center of medical care—the patient—in a way that would make the patient feel less intimidated by the medical process and help forge a stronger relation between the patient and primary care giver (we use the term "doctor" in the following summary for simplicity, understanding that many types of healthcare providers interact with patients). The aim was to use digital technology to ultimately ensure a healthier quality of life.

Technology has huge potential to empower patients to be in charge of their health. Today there are digital applications that are immediately available on smartphones and tablets to measure, monitor, consult, and track different conditions. From diagnosis to medication, data collection, computation, and data management have raised the possibility of more precise and customized healthcare. The recently launched Diabeto is an example. This is a Bluetooth device that facilitates the transfer of glucose readings from a regular gluco-meter into a specially developed Android app. The readings can then be analyzed easily with the help of the application.

But while digital technology is infiltrating medicine in newer ways at both the consumer and provider levels, some developments such as electronic medical records have pros and cons. Physicians who use electronic records during patient encounters often enter data (type) while the patient is discussing symptoms. This can be distracting and make the patient feel neglected, and even prevent him or her from discussing the full extent of their symptoms. As one participant said, "This physical examination is the most important component of diagnosis. It's not just about checking for symptoms. It's also about establishing a relationship of trust." Research has shown that "The doctor–patient relationship has been and remains a keystone of care: the medium in which data are gathered, diagnoses and plans are made, compliance is accomplished, and healing, patient activation, and support are provided."[1]

The team examined these wide ranging aspects of doctor-patient interactions and electronic medical record-keeping, but wanted to focus more on how technical innovations could further improve general wellness, rather than facilitate treatment on a case-by-case basis. The team almost unanimously hit on the idea of strengthening doctor-patient relationships as a key to making care more focused on patients, in an effort to promote personal patient responsibility in the wellness and healthcare processes.

Discussion of the critical interpersonal communication process involved in the gathering of healthcare information ultimately led to the team deciding that the terms "doctor patient relations" should refer to the patients' whole experience with the entire healthcare system starting from the time the patient enters the physician's room to hospitalization, to follow ups after discharge and overall monitoring of his health.

The Challenges

The team acknowledged the challenge of priortizing those areas that most needed change. As one participant said, "Technology can create more ways of approaching treatment but often it acts as barrier, breaking connections instead of creating them, if the doctor seems too focused on his machines instead of the patient." This is especially true in 1:1 situations

---

[1]http://www.ncbi.nlm.nih.gov/pmc/articles/PMC1496871/ (The Doctor–Patient Relationship, Challenges, Opportunities, and Strategies, Susan Dorr Goold, MD, MHSA, MA and Mack Lipkin, Jr., MD).

with patients where doctors are very hard pressed for time. Physicians, for example, can be too focused on ensuring that the patient's symptoms are entered into an electronic medical record system instead of attentively and empathetically listening to what the patient has to say.

The problem was defined as how to best design and use technology to fundamentally change the continuum of health care starting from preventive care to the diagnosis of disease to maintenance care and the reinforcement of health care, by improving the doctor–patient relationship.

A technology platform was envisioned that would help:

- Enhance the quality of the doctor-patient contact in a positive way
- Doctors and other providers change patient behaviors in a positive way for general wellness, disease prevention, or long-term or short-term health care
- Set up a system that will help the patient share responsibility for his/her wellness and health care

Exploring solutions: The big opportunity, as the team decided, is to use technical advancements to create a single 'cradle to grave' health record system that would improve doctor-patient relations by providing continuous monitoring of healthcare parameters, as well as non-intrusive care to prevent disease manage chronic conditions, and help diagnose/treat unexpected conditions like strokes and heart attacks.

Certain key parameters would be used to enhance the platform as well as ensure better health care.

These include:

- Data: Details about the patient's previous and present health, as well as lifestyle. The aim is to improve medical diagnosis based on a more comprehensive, and continuously updated personal, daily data.
- Better action and follow through: The relationship shouldn't end with the patient leaving the doctor's office. Technology must ensure that support is available to patients throughout the cycle of their care in the form of a "personal coach," tracking progress and encouraging treatment. This could occur via apps on smartphones or tablets where the patient can update data about his/her health, or apps which track and monitor the changes in their conditions.
- Information booster: Similar technologies that serve as a ready source of information or reminders for the patient at whatever point of the treatment cycle he or she is in.

Challenges with a Data Driven System

These parameters are also the points where the biggest gaps between science and technology exist. As the team discussed, comprehensive and non-intrusive data collection might seem a simple idea, but how do we know what is most relevant or what may be important in the future. Also, excess data often overwhelm and create 'noise'. Data collection here must be refined to be automated and with minimum intrusion. But data do not exist in isolation. With the introduction of a data centric platform, legal issues of privacy, regulation, and access also need to be answered. "The biggest challenges would lie," as one participant said, "in the area of validation of these data and ensuring secure access."

Another key point of discussion was preventing patients (or others) from manipulating or misusing health records. Mobile phones have been used very successfully in countries like India and South Africa to enhance health care. Would the same principles be applicable in the United States or other more developed settings?

Concrete Examples of Such Technology

The team proposed two concepts where technology can strengthen the doctor-patient relationship. While some of these solutions may already exist for use in other settings (e.g., software that can transcribe text from voice), these concepts apply exclusively to the area of health.

They included:

- "Invisible Scribe": Noting a patient's symptoms as he or she speaks is an important part of the doctor's assessment. The team envisaged a sort of digital invisible scribe system that would extract key words/phrases as the doctor and patient discussed health and wellness, and automatically organize these data into an intelligible written record of the patient encounter, thereby obviating the need for doctors to manually enter data into an electronic medical record. The system eliminates the need for the doctor having to take attention away from the patient for record-keeping, thereby enhancing the doctor-patient interaction while creating a comprehensive record of the interaction.
- "Health Ninja": This concept is designed to be a complete health care and wellness application compatible with personal digital devices of any kind to create a complete and continuously updated picture of an individual's health conditions with data collected in a non-intrusive way.

Recommendations for Research Needed

Cross-disciplinary research effort to look at the social aspects of data collection: The team knew that to make technology truly relevant to the issues they raised it would have to involve cross-disciplinary research. They recommended an approach that would bring together medical practitioners and digital technology experts, as well the disciplines of psychology and sociology to ensure that technology here would be holistic in its scope and approach.

Building a seamless patient-physician relationship: The team was clear that such diverse applications of technology, especially in the medical field, would require the development of new curricula to train a new generation of medical practitioners and healthcare technologists to help them create crucial linkages between technology and medicine. The idea isn't to create robots or let machines take over from doctors but rather enhance their interactions with patients through non-intrusive data gathering techniques.

IDR TEAM MEMBERS—GROUP B

- Fahminda N. Chowdhury, National Science Foundation
- David M. Hondula, Pavilion Research
- Amalia M. Issa, University of the Sciences in Philadelphia
- Jin Hyung Lee, Stanford University
- Ani Nenkova, University of Pennsylvania
- Desney S. Tan, Microsoft Research
- Kate Yandell, New York University

IDR TEAM SUMMARY—GROUP 6B

Kate Yandell, NAKFI Science Writing Scholar
New York University

IDR Team 6B was tasked with understanding how digital technology can be used to improve health care. The team decided to focus instead on a narrower question: How can we use digital technology to empower patients to better understand and manage their own health?

The team made this choice on the theory that doctors are knowledgeable but pressed for time. Patients and their families can be their own best advocates, because they are able to lavish the sustained attention on themselves that no one ever gets from a short doctor visit.

Let's say you have a rare cancer. Your doctor may be a top oncologist. But does she know about your particular form of cancer? Does she keep up with the latest literature? Not always. That was the case for the climate scientist Stephen Schneider, who was diagnosed with mantle cell lymphoma. He did his own literature review, assembled his own panel of experts, and convinced his doctors to try a new treatment. His cancer went into remission and he wrote about it in his book, *The Patient From Hell*.

That is an extreme case, but the team thought that perhaps all patients should be somewhat like "the patient from hell." Everyone should be engaged in his or her own care. But not everyone is a Stephen Schneider. Not everyone is even fully aware of his or her own medical history. How can we make it easier (using digital technologies) for people to keep track of their medical histories and to explore their treatment options? How can we motivate people to engage in their own care? And finally, how can we motivate people to engage in their own preventive care, such as exercising and eating healthily?

New Uses for Electronic Medical Records

The team agreed early on that patients should have easier access to their own medical data. Currently the healthcare providers who create medical records own them, but they are required by law to give patients copies upon request. The group thought that patients should have easy, electronic access to their tests results and other medical records so they could more easily monitor their own health.

Scientists would work with developers to create apps that users would be able to authorize to plug into their medical data. The team had ideas for apps for several purposes:

1. To pull out user-friendly summaries and highlights of individuals' medical data, including a monthly "health statement."
2. To aggregate medical journalism or even papers from the Web based on the user's own medical history.
3. To review users' drug prescriptions and flag potentially dangerous drug interactions or drug-food interactions.

Since users would have varying levels of medical knowledge and ability to understand complex material, the apps would be designed to provide different levels of information, depending on the users' profiles, reading habits, and ratings of the material they read.

Data would not just flow from doctors' visits. The electronic medical record would expand to include health data uploaded by the users themselves. For instance, users could sync their record to upload data from monitors recording heartbeat, blood pressure, exercise, and more. Users could also log food, mood, and sleep.

Detailed data about patients' behaviors (possibly combined with gene sequences) could help flag current problems, predict health risk, and make suggestions for lifestyle changes.

How to Help People Take Charge of Their Health and Wellbeing

Having proposed how to help people track their health, Team 6B began to wonder how many people would take the time to actually do so. The team classified people into three groups: those who are uninformed, those who are informed but unengaged, and those who are fully in charge of their own health. The team's new challenge: help people transition from unaware to aware and from aware to actively engaged.

The team decided that one of the more effective ways of getting people to engage digitally with their own health would be to focus on capturing children's interest and attention with an educational game, related to health, diet, and physical activities.

Children are often "digital natives," meaning that they grow up surrounded by technology and may be most willing and able to embrace it for learning. They are also clean slates—if we can figure out how to set positive patterns early, they may develop lifelong good habits.

The team decided the best way to teach children would be through an interactive game woven into school curriculum. Children in preschool or the early years of elementary school would play a game with a computer program or even a specially designed device that would teach them about nutrition and exercise.

They would choose avatars and be responsible for caring for their avatar's health, making meal plans and making them do physical activities.

As the avatar engaged in healthy activities, it would get health points, which would correlate with easy-to-read signs of health. (For instance, an avatar with low health points might look sluggish and unhappy.)

The games and health points would be accompanied by fun, interactive presentations on the science of nutrition and exercise. For instance, a child whose avatar had eaten broccoli might learn about the role of calcium in building strong bones.

The game would be partly theoretical, allowing the children to make any choice they wanted for the animals, but there would also be a real-world component. Children could take on special challenges in which they logged their own eating habits or recorded aerobic activities. The device could even contain a sensor that could register movement, and the child could do exercises to get points.

The culmination of the game would be the Avatar Olympics. Avatars would complete in various events, and avatars with many health points would be given advantages in the games.

The game could have a social component as well. If done in the classroom, all the students' avatars could compete in the Olympics, fostering a sense of competition and investment.

The game would, in the short run, raise children's consciousness about health and their bodies. Team members hope that in the long run it would accustom students to keeping track of their own activities and looking critically at their own lifestyle choices.

Conclusions

The digital age offers unprecedented opportunities to educate people about their own health and health care. In an era when large amounts of medical data and many choices exist, patients can, and should, become active participants in their own medical care decisions.

Team 6B's apps for electronic medical records would help adults access and, more important, interpret information about their own bodies. Engaged, empowered patients would take better care of themselves both within and outside the doctor's office. One of the great challenges of medicine is to improve preventive care. Digital tools are ideal for improving preventive care, as preventive care happens when patients are outside the medical system.

Today, it may be hard to imagine many people beyond extreme health enthusiasts logging their health data and playing with health apps. However, Team 6B thinks that a new generation, at home in the digital world and engaged early in life, has the potential to take responsibility for their health.

IDR Team Summary 7

What are the limits of the Brain-Computer Interface (BCI) and how can we create reliable systems based on this connection?

CHALLENGE SUMMARY

BCI includes a wide range of interface and signal processing technologies from direct recordings from brains to electroencephalography (EEG) and functional magnetic resonance imaging fMRI. BCI enables a wide range of applications that include helping those with impaired physical function, such as stroke victims, control everyday objects in their environment; analyzing awake and sleep brain states to monitor alertness levels and diagnose brain disorders, and understanding market preferences.

One of the most dramatic advances in recent years is "mind reading," which uses BCI to decode brain states to reconstruct what a subject is experiencing. There is also a growing market for computer game and devices that are controlled by brain states (http://www.bcireview.com/). Wireless technology has made it possible to record from mobile humans.

Another active area of BCI is replacement of lost sensory interfaces. Cochlear implants were developed in the 1970s and over 219,000 people worldwide have received cochlear implants. Progress has also been made on retinal and cortical implants to restore sight in blind patients. Remarkably, blind patients have reported substantial "sight" using a camera to activate an array of electrodes on the tongue, one of the most sensitive sensory surfaces of the body.

Key Questions

- What are the technical problems with creating long-term, stable interfaces with brains?
- Can two humans implanted with BCI communicate directly with each other? What would be the consequences?
- There are many ethical concerns including: Consent, privacy re. mind reading, fear of hype, personality alteration, risks and benefit. As with other powerful technologies, BCI can be used for good and bad purposes. What impact will it have on society?

Reading

Berger TW and Glanzman DL (Eds.). Toward replacement parts for the brain: Implantable biomimetic electronics as neural prostheses. MIT Press: Cambridge, MA, 2005.

Makeig S, Gramann K, Jung T-P, Sejnowski TJ, Poizner H. Linking brain, mind and behavior. *Int J Psychophysiol* 2009;73: 95-100.

Nishimoto S, Vu AT, Naselaris T, Benjamini Y, Yu B, Gallant JL. Reconstructing visual experiences from brain activity evoked by natural movies. Curr Biol 11 Oct 2011;21(19):1641-6. Epub 2011 Sep 22.

Because of the popularity of this topic, three groups explored this subject. Please be sure to review each write-up, which immediately follow this one.

IDR TEAM MEMBERS—GROUP A

- Paul A. Fishwick, The University of Texas at Dallas
- Joseph T. Francis, SUNY Downstate Medical School
- Nikki Mirghafori, International Computer Science Institute at Berkeley
- Jacquelyn F. Morie, University of Southern California
- Adriane B. Randolph, Kennesaw State University
- Ravishankar Rao, IBM Research
- Aviva Hope Rutkin, Massachusetts Institute of Technology
- Paul Sajda, Columbia University

IDR TEAM SUMMARY—GROUP 7A

Aviva Hope Rutkin: NAKFI Science Writing Scholar
Massachusetts Institute of Technology

IDR Team 7A was asked to explore the possibilities and limitations of brain-computer interface (BCI) technology. Brain-computer interfaces are devices that permit direct communication between the brain and an external machine. Such communication could involve tracking an individual's mental state, altering the way the brain processes information, or even affecting neurological operations in real time.

Team 7A was fascinated by the many potential products that could flood a future BCI market. They opted to act as the board of an imaginary company called Brain Buddy, Inc. (Motto: "We help you think.") By brainstorming numerous possible devices to push into development, the team was able to explore by proxy the potential applications and pitfalls of BCI technology.

Guiding Considerations

First, the team considered some larger issues for its discussion. These questions, which they divided into the categories shown below, kept the members attuned to the many current constraints of BCI technology.

Science

Our group had to consider the limits of human knowledge. There is much that we do not yet understand about the way the brain functions. We also do not understand the relationship between consciousness and subconsciousness. For many of Brain Buddy's proposed products, the team looked for previous research or ongoing experimentation to suggest that a marketable BCI product could one day be possible.

Ethics

Introducing new technology without deeply considering its impact on the world could have disastrous consequences. For example, some people might become physically or cognitively dependent on their BCI device. Though issues of social policy were a recurring theme throughout the conversation, some of the team alluded that it is up to society to decide if a given kind of technology is acceptable or unacceptable. As one member put it, "the street finds its own use." All agreed to the value of the conversation.

Law

BCI technology raises many new ethical questions that will have to be answered in our legislatures and courts. Who owns the copious data generated by, e.g., a "lifelogging" device that continuously records your biological statistics? How can we ensure privacy of BCI devices? How do we prevent others from hijacking our minds for mischievous or even nefarious purposes? The intense public concern about President Barack Obama's Blackberry in 2008 pales in comparison to the long-term possibility that terrorists could hack into his brain.

Control

The existence of many of Brain Buddy's devices is predicated on the development of quite a few new engineering techniques, including the ability to handle sensory input and motor output. Many devices would also require fully-closed loop BCI bi-directionality.

Prototype Development

With these issues in mind, the team explored the rich and varied world of future BCI.

The team began by focusing on a single imaginary product: a labeling cap. This cap would monitor the brain for identifying signals, such as the P300, a well-researched electric signal that indicates when attention has been piqued. This signal would trigger the cap to take a picture of your surroundings and tag it with other important data like time and GPS location. The picture would then be automatically uploaded to the cloud, where a powerful algorithm would sort the picture into a category with others.

They debated possible users of this technology. An ordinary busy person like an academic or an artist might use the pictures to generate new ideas for their work. Someone with Alzheimer's could use the cap as a kind of memory supplement. Others might use it to create a visual diary of their life.

One problem is that this device would generate a massive amount of data—far too much for any rational person to sift through on a regular basis. The group agreed that very powerful computer algorithms would have to learn what information you want to keep or discard. This opened up the possibility for the computer to use the data to provide just-in-time feedback. Instead of manually sorting your own data, computer scientists could develop programs that deliver automatic notifications, or even provide unprompted direct brain stimulation.

Team 7A also considered ways in which other people might benefit from a labeling cap. If pertinent mental, emotional, and physiological data could be outwardly displayed in, say, a projection directly above our heads, how would that impact the way in which we interact with others? Such technology could ultimately lead to the kind of mind-reading "which uses BCI to decode brain states to reconstruct what the subject is experiencing" suggested in NAKFI's prompt for IDR Team 7. (However, some members suggested that this kind of mind meld would be too high-bandwidth for humans to handle.)

Proposed Products

The label cap was a launchpad for a long list of new Brain Buddy products. These included:

Brain Buddy (standard model)

Many individuals wish that they were more productive. However, we often get bogged down in time-wasting tasks and are easily drawn by tantalizing distractions. Now, by wearing the standard Brain Buddy, you will be notified about how to best portion your time. The device will track what settings tend to lead to your best work, and provide suggestions about how to spend your workday. It will also notify you through a gentle tone when to take a break from work, preventing you from becoming excessively stressed.

Health-A-Wear

Imagine that a veteran with post-traumatic stress disorder (PTSD) needs to check her mailbox. The neighbor's sprinklers unexpectedly turn on as she walks down the driveway, triggering a panicked emotional response. The Health-A-Wear device would be primed to handle these issues, perhaps through noise cancellation, post-stimulus masking, or even direct-brain stimulation. It could learn which settings were more likely to cause problems and prepare the individual accordingly. Over time, the device would slowly decrease its functioning, allowing the user to wean herself off the BCI.

Learn-A-Wear

In an online course, the professor has little to no understanding of how well the students are learning the material. However, if all of the students were wearing Learn-A-Wear devices, he would be able to monitor each of their mental states. This would allow him to calibrate his lesson to the advantage of each individual listener, as well as provide meaningful feedback on his overall performance. The students' caps could also filter an incoming presentation for information that is most pertinent or helpful to them.

Brain Buddy, Jr.

When a baby cries, his parents must guess at what's troubling him. Is he hungry? Does he need his diaper changed? Does he need to be burped? Now, by outfitting your baby with Brain Buddy, Jr., you can instantly check your baby's state and understand what is on his mind. Furthermore, software upgrades would become available as your child aged. When he reaches, for example, the often problematic middle school years, you would be able to monitor his emotional state for signs of anxiety or depression.

Brain Buddy Silver

As you age, it can be difficult to keep track of upcoming health problems. Brain Buddy Silver would be calibrated to track your body for signs of possible medical issues, catching problems before they became dire. It would also monitor your mental state, suggesting, e.g., memory games to keep you alert and boost your cognitive functioning.

Additional Issues

Team 7A agreed that they were interested in developing some of the proposed products. While fine-tuning the details of the different prototypes, they encountered several more problems that needed to be addressed by their hypothetical company. In order to be successful, Brain Buddy would have to find ways of:

- making the product ergonomic and aesthetically pleasing
- ensuring that the product met some minimum threshold of reliability, as a technological malfunction could be troublesome or even dangerous
- developing an efficient algorithm to retrieve, store, integrate, and secure neurological data
- keeping the device relatively affordable
- finding a practical way to power the device
- calibrating a single device to different brains and thinking styles

Furthermore, ethical questions continued to loom large. It was not difficult to imagine deeply troubling scenarios in which subversive groups reverse-engineered BCI technology to conduct cyber-neuronal warfare.

Team 7A was also particularly concerned with the possible adverse health effects that could result from BCI use. For example, some users might become dependent on or addicted to their BCI; one member suggested that a lawsuit would result if one person took another somewhere out of range of wireless connection. Other users might develop "data-compulsive disorder," becoming obsessed with the experience of lifelogging. Some could lose touch entirely with the real world.

Many of these issues came back to one large, overarching problem: we still don't know how the brain works. We don't know how wearing a BCI device would impact our nervous system. We don't know if these unpredictable plasticity changes would be a positive or negative. Though the team ended its discussion feeling optimistic about Brain Buddy's potential, they agreed that many questions must be answered to make BCI technology a reality.

IDR TEAM MEMBERS—GROUP B

- Dima Amso, Brown University
- Cynthia S. Atherton, Gordon and Betty Moore Foundation
- Jose M. Carmena, University of California, Berkeley
- John Doyle, California Institute of Technology
- Adam Gazzaley, University of California, San Francisco
- Ricardo Gil da Costa, Salk Institute
- Jay Lee, University of Cincinnati
- Chris Palmer, University of California, Santa Cruz
- Anna W. Roe, Vanderbilt University
- Aaron L. Williams, University of Virginia

IDR TEAM SUMMARY—GROUP 7B

Chris Palmer, NAKFI Science Writing Scholar
University of California, Santa Cruz

IDR Team 7B was asked to define the limits of Brain-Computer Interfaces (BCI) and determine the reliability of systems based on this connection.

One of the most active fields within basic biological sciences over the past 20 years has been brain science. The variety of methods to record brain activity is rapidly growing. Also, the spatial and temporal resolution of brain activity signals is improving, meaning it is easier to tell when a precise area of the brain is activated. At the same time, advances in electronics and computing have led to the miniaturization of robust and powerful computing devices. This convergence of technological innovation is making it easier to bring computers into close contact with brains to acquire reliable brain activity signals and control brain activity in meaningful ways.

The team quickly decided it did not want to focus on the well-trod topic of technical engineering challenges related to improving BCI. Instead, the team turned it attention to how computers can work with our brains to enhance cognitive ability.

Inspired by Clifford Nass' plenary talk, the team focused on one cognitive function that is relatively poor in humans: directing attention. As multitasking increasingly becomes a part of our everyday experience, it becomes difficult to know what one should be paying attention to from moment to moment. Humans can benefit from an ongoing stream of computer-generated cues about which features in the environment to attend to. The team also emphasized a technological approach that maintains or even enhances face-to-face social contact, again inspired by issues raised in Nass' talk. Nass, a professor of communication at Stanford University, presented experimental results showing that heavy multitaskers were actually poor at multitasking due to a deteriorated ability to focus attention. He also emphasized the necessity of direct social interactions for developing healthy emotional responses.

The Ultimate Brain-Computer Interface for the Digital World

Our team envisioned a closed loop BCI device that assesses a person's environment, her life history and goals and her current brain and body states, and uses this information to influence her digital environment. The BCI would then provide an input to the brain's attention centers to focus attention in the desired manner.

Input signals to the BCI

There are a variety of real time signals about a person that can be fed into the BCI, such as electrical signals from the brain and physiological signals from the body (pulse, respiration, eye-movements, facial expression, etc.). BCI can also form an overall history for its user based on weeks and months of collecting real time data. Medical history can also be incorporated.

Types of BCI

The team did not specify what kind of BCI technology would read brain signals or stimulate attention areas. Some of the non-invasive options for reading brain signals include EEG and functional imaging. Invasive options include direct microelectrode array recordings and optical imaging.

Non-invasive options for brain stimulation include transcranial magnetic stimulation and transcranial electrical stimulation. Invasive options include optogenetics and direct microelectrode array stimulators.

How Does the Brain Focus and Maintain Attention?

Two aspects of attention

There are two primary aspects of attention that can be guided by the BCI. Selective attention indicates where, or to what, attention is directed. Sustained attention indicates a continued holding of attention at the selected location or object. The former can be manipulated to facilitate task switching and the latter can be manipulated to facilitate task-maintenance.

Brains areas involved in attention

Neurophysiological studies in monkeys show that electrical stimulation of visual cortical areas with microelectrodes is effective at directing visual attention to certain locations in the environment. Because visual cortex contains map-like representations of the environment, it is relatively straightforward to figure out which small portion of the visual cortex to stimulate to selectively focus visual attention at a specific spatial location.

Like the maps in visual cortex, many brain areas contain spatially arranged maps of various features that have been well studied by neuroscientists. These areas can be similarly targeted with electrical stimulation to selectively focus attention on any number of specific environmental features. For example, precise stimulation of auditory cortex could bring attention to sounds of specific frequencies or tones—making it possible for the BCI to direct a person's attention to a particular voice or a sound that conveys an imminent threat.

This type of stimulation could be very helpful in a number of clinical populations. For example: 1) Individuals with autism often have difficulty reading emotion and may benefit from stimulation to brains areas that process humans faces when they engage in social interactions; 2) Victims of stroke or other brain damage may experience hemifield neglect, in which they ignore one half of their body and the external world. Directed stimulation to the sensory cortex, which contains map-like representations of the body, can alert these individuals in cases where there is risk to a neglected body part—e.g., it is about to come in contact with a hot stove or sharp object; and 3) For those with emotional disorders, limbic structures may be stimulated to modulate positive and negative reactions to specific events or stimuli.

The above are examples of how a BCI could use electrical stimulation to direct selective attention. Frontal and parietal cortical areas in the right hemisphere, as well as clusters of brain stem nuclei, have been shown to be important for sustaining attention in certain behavioral tasks. Electrical stimulation of some collection of these areas could assist people with maintaining attention on important tasks.

Enhancing Attention

The algorithm used by the BCI can be tuned, or designed, to fit the needs or desires of an individual. Specialized algorithms may be made for individuals engaged in specific tasks. Other algorithms may be specialized for special needs populations.

Education

Giving children cues about where to focus their attention (selective attention) while engaged in learning something, as well as helping them sustain that attention, can accelerate the learning.

On the flip side, perseveration is an inability to disengage attention. Children who perseverate on an object or task could benefit from a BCI that quickly recognizes the behavior, as well as neural and brain signals that proceed the behavior, and direct the child's attention to a new object or task.

Enhancement

BCI can be used to most efficiently direct attention during multi-tasking sessions. In a case where an individual has a primary and secondary task, the individual must occasionally break away from the primary task to perform the secondary task. The problem is that people are bad at knowing when is the best time to make these switches. Task switching during a period of sustained concentration can be counter productive. The BCI can detect oncoming, naturally occurring dips in attention during the performance of the primary task and induce a switch to the secondary task.

Therapeutics

There are many disorders of attention that can be addressed with the proposed BCI system, including autism, mood disorders, and brain injury.

Conclusions

Team 7B designed a general framework for a human-centered BCI that can be useful for education, enhancing every day life and providing therapeutic interventions. Though we focused on one specific cognitive function, i.e. attention, the framework can tackle additional functions such as perception, motor control, memory, language, etc.

Though the team did not discuss specifics about when BCI systems to control cognitive functions may be ready for testing in humans, most of the system components already exist in some form. Over the past few years, flexible electrode arrays have been implanted in human patients to detect the onset of epileptic seizures. Similar technology could be implanted within the brain areas discussed earlier to control some aspects of attention via electrical stimulation.

Brain stimulation is also possible via non-invasive technology such as transcranial magnetic stimulation and transcranial direct current stimulation. Here, stimulation is delivered with electrodes placed on the scalp above the target brain area. However, further advances are needed to provide stimulation to precise locations in the brain.

IDR TEAM MEMBERS—GROUP C

- Todd P. Coleman, University of California, San Diego
- Vincent DeSapio, HRL Laboratories, LLC
- Satinderpall S. Pannu, Lawrence Livermore National Laboratory
- Thomas Serre, Brown University
- Kelly Servick, University of California, Santa Cruz
- Qi Wang, Georgia Institute of Technology/Emory University
- Byron M. Yu, Carnegie Mellon University

IDR TEAM SUMMARY—GROUP 7C

Kelly Servick, NAKFI Science Writing Scholar
University of California, Santa Cruz

IDR Team 7C was asked to probe the limits of the Brain-Computer Interface (BCI) and to suggest how we might create reliable systems based on this connection. Because team members had a rich collective expertise in restorative systems (seeking to restore lost sensory or motor abilities in a clinical setting), early discussion focused on the state of the art and the technical obstacles to effective motor control through BCI.

To explore the outer limits of current technology for decoding information from brains, the team tried to envision scenarios of direct brain-to-brain communication. Potential benefits of such communication range from a new means of self-expression for "locked-in" patients to non-linguistic communication among all humans. This technology might even enhance our social lives in a digital age where communication is increasingly carried out in an emotionally restrictive online environment.

However, questions of implementation quickly overshadowed the theoretical discussion. The extreme challenges of decoding neural signals into relevant information, maintaining the integrity of implanted interfaces over time, and generalizing among unique, individual brains all came to the fore.

The team struggled to address more abstract questions about information transfer without returning to the field's current limitations.

In the end, the technical discussion laid a meaningful groundwork for developing a more creative application of BCI. The team put aside technical specificity, but retained the somewhat utilitarian spirit of BCI, to develop the forward-looking concept of "brainlogging" as a means of enhancing our shared digital experience.

Lessons from Neural Prosthetics: Two Fundamental Dichotomies

The field of BCI faces significant trade-offs between methodological alternatives. First is the question of invasive versus non-invasive technology. The team agreed that much information can be gathered from outside the brain, without the surgical implantation of electronics, but that such information can serve only very specific, limited purposes. The skull is ultimately a powerful insulator of signals, and the only way to record or stimulate precise neurons or neural populations is by opening the skull and interacting directly with brain tissue. However, such a radical procedure has limited potential for use in humans, particularly those not seeking solutions to a severe physical disability. In fact, even among amputees and paralyzed individuals, resistance to invasive BCI is common.

A second source of tension in the field concerns human physiology. There is debate about whether knowledge of the brain is a necessary component of BCI from an engineering perspective. For example, in an algorithm that transforms neural activity into the movement of a robotic arm, programmers need not understand the functional role of individual neurons or small groups of neurons. Most of the signals the neurons produce are collapsed or discarded in the process of translating brain data into meaningful information, like instructions for a robotic arm. However, the team also acknowledged that a better understanding of the neural physiology that underlies interrelated systems in the brain might lead to better computational models and more effective BCI.

Though these issues could not be resolved in the course of a conference, neither were they ignored. The question of invasiveness highlighted the need for considering the desires of the end user. The tension between brain physiology and device functionality expressed the many levels on which the brain can by explored and mined for useful information. Both questions would inform the evolving discussion.

Asking a New Question

Knowing that the limitations of current technology and computational modeling are constantly changing, the team struggled to find a question that would set these issues aside. The goal was to find an application for BCI that might have desirable effects for humans in the digital age, regardless of the technological platform on which it was implemented. Assuming that brain monitoring could gather large datasets, what information would we want? The team's question was refined to: what information can we draw from neural activity that we're not already measuring directly in some other way?

This line of thought led to the concept of lifelogging—the focus of IDR Team 3. Currently it is possible to record all kinds of data from digital users (GPS location, heart rate, sleep patterns, jogging speed, caloric intake, etc.) It is also possible to log our opinions through built-in features of the digital environment: the "like" button on Facebook and numerical rating systems like awarding "stars" to a Netflix movie or an Amazon product are examples. However, conveying subtle emotional reactions in the digital world can be challenging. By nature of their brevity, comment threads and status updates rarely contain rich, emotionally introspective content.

By monitoring brain activity, Internet users could record their complex mental states in real time, bypassing the task of formulating and broadcasting emotions as text.

These brain logs could be shared and compared, allowing a user to create a personalized online "signature" and match it with the signatures of other users. Two people who exhibit similar neural responses to a piece of online media might have other meaningful similarities. (Work by Hasson et al. in 2004 has already shown that all viewers have closely synchronized neural activity while watching the same movie.) A process that the group dubbed "brainlogging" could focus on subtle similarities and differences in brain activity to connect users who have parallel emotional responses to the same online experience.

Fleshing Out the "Brainlogging" Concept

The group envisioned a system that records and stores neural activity while a person navigates the Internet and interacts with digital media. They chose not to specify what type of brain imaging technology would be used, or which neural regions would be monitored, judging these technical considerations to be outside the scope of the meeting. However, it was assumed for purposes of discussion that scientists could already chronically record neural activity for large populations using minimally invasive, low-cost technology—a scenario they agreed was futuristic but not unrealistic. The remaining challenges in creating such a system would be:

- Analyzing huge sets of brain data to extract the interesting or relevant features
- Timestamping online media so that each neural response could be connected to the content that provoked it
- Determining which brain activity is relevant, and if necessary, separating purely sensory functions (eye movements to track video, for example), from deeper emotional responses

Even if innovative computer science models could address these issues, broader concerns remain. This extreme form of datalogging raises questions of privacy, particularly if such data were incorporated into social media. Companies might be motivated to use brain data to target advertising more directly, capitalizing on subconscious emotional characteristics of users to manipulate buying behavior. A user's data might also be used as a predictive tool to draw conclusions about character flaws or even criminal tendencies. Employers might use such data to discriminate among candidates based on their neural habits. As with any personal information collected and evaluated in the context of the Internet, questions of social policy abound.

Envisioning a "Brainlogged" Future

Beyond the originally envisioned benefits of neural recording, the team identified other parts of our lives that might be revolutionized in a "brainlogged" environment.

Users might reap health benefits from having memory banks to store neural data. Medical professionals could analyze these data as another diagnostic tool or as a way of monitoring patient wellbeing. The system might be able to identify the neural precursors to certain illness. In particular, the team suggested that diseases like Alzheimer's and Parkinsons have distinct neural "warning signs" that might enable early detection. Individuals with mental illness such as depression might benefit from more thorough tracking of their emotional states.

The system also has potential as an educational tool. Online learning environments often lack sufficient feedback about student understanding, attentiveness, and investment. This direct form of monitoring could allow educators to track the neural patterns of their students closely and adjust the learning environment to help students succeed.

Finally, an application that might resonate in an age of increasing social fragmentation is "bHarmony"—a romantic matching system based on neural similarities and shared emotional responses. (Perhaps neural logging would offer more personality insight than a series of multiple-choice questions. . . .) The group took into account evidence presented at this year's NAKFI conference about the possible decline in meaningful human interaction in the digital age. The members suggest that "brainlogging" technology could someday create a healthier, more emotionally enriching world around our changing brains.

Appendixes

List of The Informed Brain in a Digital World Podcast Tutorials

The Effects of the Digital Age on Education
Podcast Released: September 13, 2012
Michael Keller
Ida M. Green University Librarian
Director of Academic Resources
Stanford University

The Trajectory, Value, and Risk of Extreme Life Logging
Podcast Released: September 20, 2012
Cathal Gurrin
SFI Stokes Lecturer School of Computing
University of Ireland, Dublin City

The Impact of the Internet on the Social Behavior
Podcast Released: September 27, 2012
Robert Kraut
Herbert A. Simon Professor of Human-Computer Interaction
Human-Computer Interaction Institute
Carnegie Mellon University

Differences in Cognitive and Brain Function of Digital Natives and Digital Immigrants
Podcast Released: October 2, 2012
Clifford Nass
Thomas M. Storke Professor of Communication
Stanford University

The Effects of the Digital Age on Health and Wellness
Podcast Released: October 11, 2012
Larry Smarr
Director, Calit2
University of California, San Diego

The Brain-Computer Interface
Podcast Released: October 17, 2012
Scott Grafton
Professor
Psychological & Brain Sciences
University of California, Santa Barbara

All tutorials are available at www.keckfutures.org.

Agenda

Thursday, November 15, 2012

8:00 a.m.	Bus Pickup: Attendees are asked to allow ample time for breakfast at the Beckman Center; no food or drinks are allowed in the auditorium, which is where the welcome and opening remarks take place at 9:30.
8:30 a.m.	Registration (not necessary for individuals who attended the Welcome Reception)
8:30—9:30 a.m.	Breakfast
9:30—9:45 a.m.	**Welcome and Opening Remarks** Harvey V. Fineberg, President, Institute of Medicine Michael S. Gazzaniga, Chair, NAKFI Steering Committee on the Informed Brain
9:45—10:45 a.m.	**Keynote Address** Patrick Suppes (NAS), Professor Emeritus, Stanford University
10:45–11:00 a.m.	**Interdisciplinary Research (IDR) Team Challenge and Grant Program Overview** Michael S. Gazzaniga, Chair, NAKFI Steering Committee on the Informed Brain

11:00—11:15 a.m.	Break
	Poster Session A Setup
11:15 a.m.—12:45 pm	Poster Session A
11:15 a.m.	Graduate Students meet with Barbara Culliton
12:45—2:00 p.m.	Lunch
2:00—5:30 p.m.	IDR Team Challenge Session 1
3:00—3:30 p.m.	Break
	Poster Session B Setup
5:30—7:00 p.m.	Reception/Poster Session B
7:00 p.m.	Bus Pickup: Attendees brought back to hotel

Friday, November 16, 2012

8:00 a.m.	Bus Pickup
8:15—9:00 a.m.	Breakfast
9:00—11:00 a.m.	IDR Team Challenge Session 2
11:00—11:30 a.m.	Break
11:30 a.m.—1:00 p.m.	IDR Team Challenge Preliminary Reports (5 to 6 minutes per group)
1:00—2:00 p.m.	Lunch
2:00—5:30 p.m.	IDR Team Challenge Session 3
3:00—3:30 p.m.	Break
5:30 p.m.	IDR Team Challenge Final Presentation Drop-Off: IDR Teams to drop off presentations at information/registration desk, or upload to FTP site prior to 7:00 a.m. Saturday morning
5:45 p.m.	Bus Pickup: Attendees brought back to hotel for a free night

Saturday, November 17, 2012

7:00 a.m.	Bus Pickup: Attendees who are departing for the airport directly from the Beckman Center are asked to bring their luggage to the Beckman Center. Storage space is available.
7:15—8:00 a.m.	Breakfast
7:15 a.m.	Taxi Reservations: Attendees are asked to stop by the information/registration desk to confirm their transportation to the airport or hotel.
8:00—9:30 a.m.	IDR Team Challenge Final Reports (8 to 10 minutes per group)
9:30—10:00 a.m.	Break
10:00—noon	IDR Team Challenge Final Reports (continued) (8 to 10 minutes per group)
11:00 a.m.—noon	Q&A Across All IDR Teams
Noon—1:30 p.m.	Lunch (optional)

Participant List

Dima Amso
Assistant Professor
Cognitive, Linguistic, &
 Psychological Science
Brown University

Giorgio A. Ascoli
University Professor
Molecular Neuroscience
George Mason University

Paul Atchley
Professor
Psychology
University of Kansas

Cynthia S. Atherton
Program Director
Science Program
Gordon and Betty Moore
 Foundation

Lisa Aziz-Zadeh
Assistant Professor
Brain and Creativity Institute &
 Division of Occupational
 Science
University of Southern California

David Badre
Assistant Professor
Cognitive, Linguistic, and
 Psychological Science
Brown University

Richard G. Baraniuk
Victor E. Cameron Professor
Electrical and Computer
 Engineering
Rice University

Carole R. Beal
Professor of Information Science
 Technology & Arts
School of Information Science
University of Arizona

C. Gordon Bell
Principal Researcher
Microsoft Research

Robert M. Bilder
Tennenbaum Family Chair of
    Creativity Research
Department of Psychiatry &
    Biobehavioral Science
UCLA Semel Institute for
    Neuroscience & Human
    Behavior

Kim T. Blackwell
Professor
Molecular Neuroscience
George Mason University

Alison Bruzek
Science Writing Scholar
Graduate Program in Science
    Writing
Massachusetts Institute of
    Technology

Jose M. Carmena
Associate Professor
Electrical Engineering &
    Computer Sciences;
    Neuroscience
University of California, Berkeley

Tara T. Cataldo
Biological/Life Sciences Librarian
Marston Science Library
University of Florida

Vinton G. Cerf
Chief Internet Evangelist
Research
Google, Inc.

Fahmida N. Chowdhury
Program Director
Social, Behavioral and Economic
    Sciences
National Science Foundation

Ann E. Christiano
Frank Karel Endowed
    Chair in Public Interest
    Communications and
    Professor
College of Journalism and
    Communications
University of Florida

John-Paul Clarke
Associate Professor
Aerospace Engineering
Georgia Institute of Technology

Mark Steven Cohen
Professor
Psychiatry, Neurology, Radiology,
    Psychology, Biomedical
    Engineering, Biomedical
    Physics
University of California, Los
    Angeles

Todd P. Coleman
Associate Professor
Bioengineering
University of California, San Diego

PARTICIPANTS

David C. Cook
Senior Economist
Department of Agriculture and Food
Government of Western Australia

Alan DJ Cooke
Associate Professor
Marketing
University of Florida

Jana Craig Hare
Assistant Research Professor
Center for Research on Learning
University of Kansas

Carolyn Crist
Science Writing Scholar
Grady College of Journalism and Mass Communication
University of Georgia

Barbara J. Culliton
President
The Culliton Group/Editorial Strategies

Robert J. Davenport
Associate Director
Brown Institute for Brain Science
Brown University

Vincent De Sapio
Research Scientist
Information and Systems Sciences Laboratory
HRL Laboratories, LLC

John Doyle
Professor
Control and Dynamical Systems
California Institute of Technology

Nicole B. Ellison
Associate Professor
School of Information
University of Michigan

Harvey V. Fineberg
President
Institute of Medicine

Paul A. Fishwick
Professor
Computer and Information Science & Engineering
University of Florida

Dorothy Fleisher
Program Director
W.M. Keck Foundation

Roxanne Ford
Program Director
W.M. Keck Foundation

Chris Forsythe
Distinguished Member of Technical Staff
Cognitive Science and Applications
Sandia National Laboratories

Richard N. Foster
Lux Capital
Yale University

Joseph T. Francis
Assistant Professor
Physiology and Pharmacology
SUNY Downstate Medical School

Felice C. Frankel
Research Scientist
Center for Materials Science and
    Engineering
Massachusetts Institute of
    Technology

Kenneth R. Fulton
Executive Director
National Academy of Sciences

Adam Gazzaley
Associate Professor
Neurology, Psychiatry and Physiology
University of California, San
    Francisco

Michael S. Gazzaniga
Director
The Sage Center for the Study of
    the Mind
University of California, Santa
    Barbara

Apostolos Georgopoulos
Regents Professor and McKnight
    Presidential Chair in Cognitive
    Neuroscience
Brain Sciences Center
American Legion Brain
    Sciences Chair; Professor of
    Neuroscience, Neurology, and
    Psychiatry
Veteran Affairs Medical Center

Ricardo Gil da Costa
Salk Institute

Charles D. Gilbert
Arthur and Janet Ross Professor
Laboratory of Neurobiology
The Rockefeller University

Arnold L. Glass
Professor
Psychology
Rutgers University

Scott T. Grafton
Professor
Psychological & Brain Sciences
University of California, Santa
    Barbara

Robert A. Greenes
Ira A. Fulton Chair and Professor
Department of Biomedical
    Informatics
Arizona State University

Cathal Gurrin
SFI Stokes Lecturer
School of Computing
Dublin City University

David S. Hachen
Associate Professor
Sociology
University of Notre Dame

Todd F. Heatherton
Lincoln Filene Professor in Human
    Relations
Dartmouth College

Anne Heberger Marino
Sr. Program Associate
National Academies Keck *Futures Initiative*

Matthew K. Henley
Ph.D. Candidate
Educational Psychology
University of Washington

Zahra Hirji
Science Writing Scholar
Graduate Program in Science Writing
Massachusetts Institute of Technology

David M. Hondula
Research Staff
Department of Environmental Sciences
University of Virginia

Amalia M. Issa
Professor and Chair and Director
Health Policy and Public Health
Program in Personalized Medicine & Targeted Therapeutics
University of the Sciences in Philadelphia

Michael A. Keller
Ida M. Green University Librarian; Director of Academic Information Resources
Cecil H. Green Library
Stanford University

Cristen A. Kelly
Associate Program Officer
National Academies Keck *Futures Initiative*

Steven Kotler
Director of Research
Flow Genome Project

Art Kramer
Director and Professor
Beckman Institute
University of Illinois

Shonali Laha
Associate Professor
Civil and Environmental Engineering
Florida International University

Annie Lang
Distinguished Professor
Telecommunications, Cognitive Science
Indiana University

Kenneth M. Langa
Professor of Medicine
Internal Medicine and Institute for Social Research
University of Michigan

Eva K. Lee
Director, NSF-Whitaker Center
 for Operations Research in
 Medicine and HealthCare;
 Co-Director, NSF-I/
 UCRC Center for Health
 Organization Transformation;
 Professor, School of Industrial
 & Systems Engineering;
 Distinguished Scholar in
 Health
Georgia Institute of Technology

Jay Lee
Ohio Eminent Scholar and L.W.
 Scott Alter Chair Professor
School of Dynamic Systems
University of Cincinnati

Jin Hyung Lee
Assistant Professor
Neurology and Neurological
 Sciences, Bioengineering
Stanford University

Kalev H. Leetaru
Josie B. Houchens and University
 of Illinois Fellow
Graduate School of Library and
 Information Science
University of Illinois

Mark W. Lenox
Director of Imaging
Imaging Core
Texas A&M Institute for
 Preclinical Studies

Rachel Lesinski
Program Associate
National Academies Keck *Futures
 Initiative*

Jeffrey Liew
Associate Professor
Educational Psychology
Texas A&M University

John Linehan
Professor of Biomedical
 Engineering
Biomedical Engineering
Northwestern University

Julie Linsey
Assistant Professor
Woodruff School of Mechanical
 Engineering
Georgia Institute of Technology

Taosheng Liu
Assistant Professor of Psychology
Psychology
Michigan State University

Ning Lu
Associate Professor
Electrical and Computer
 Engineering
North Carolina State University

Wei Lu
Associate Professor
Mechanical Engineering
University of Michigan

Jessica Luton
Science Writing Scholar
Grady College of Journalism and
    Mass Communication
University of Georgia

Margaret Y. Mahan
Graduate Student
Biomedical Informatics and
    Computational Biology
University of Minnesota

Andreas Malikopoulos
Alvin M. Weinberg Fellow
Energy & Transportation Science
    Division
Oak Ridge National Laboratory

Narayan B. Mandayam
Peter D. Cherasia Faculty Scholar
    and Professor of ECE
Electrical and Computer
    Engineering
Rutgers University

Gloria Mark
Professor
Department of Informatics,
    Interactive and Collaborative
    Technologies
Donald Bren School of Information
    and Computer Sciences
University of California, Irvine

Dejan Markovic
Associate Professor
Electrical Engineering
University of California,
    Los Angeles

René Marois
Professor
Psychology
Vanderbilt University

Ulrich Mayr
Professor
Psychology
University of Oregon

Kirk McAlpin
Freelance Science Writer

John Devin McAuley
Associate Professor/Director,
    Interdisciplinary Cognitive
    Science Program
Psychology
Michigan State University

Todd J. McCallum
Associate Professor of Psychology
Psychological Sciences
Case Western Reserve University

David E. Meyer
Clyde Coombs Distinguished
    University Professor of
    Mathematical Psychology
Psychology/Cognition and
    Cognitive Neuroscience
University of Michigan

Nikki Mirghafori
Senior Research Scientist
International Computer Science
    Institute at Berkeley

Jacquelyn F. Morie
Senior Research Scientist
Institute for Creative Technologies
University of Southern California

Hamid Najib
Information Technology and
    Program Support Specialist
National Academies Keck *Futures
    Initiative*

Clifford Nass
Thomas M. Storke Professor
Communication
Stanford University

Ani Nenkova
Assistant Professor
Computer and Information
    Science
University of Pennsylvania

Oded Nov
Assistant Professor
New York University–Polytechnic
    Institute

Meeko Mitsuko Karen Oishi
Assistant Professor
Electrical and Computer
    Engineering
University of New Mexico

Anthony C. Olcott
Scholar
Emerging Trends—Center for the
    Study of Intelligence
Central Intelligence Agency

Paromita Pain
Science Writing Scholar
Annenberg School for
    Communication & Journalism
University of Southern California

Chris Palmer
Science Writing Scholar
University of California, Santa
    Cruz

Satinderpall S. Pannu
Section Leader for the Center for
    Micro- and Nano-Technology
Engineering
Lawrence Livermore National
    Laboratory

Roy Pea
David Jacks Professor of Education
    and Learning Sciences;
    Director, H-STAR Institute
Graduate School of Education/H-
    STAR Institute
Stanford University

Maria Pellegrini
Executive Director, Programs
W.M. Keck Foundation

Kimberly F. Raab-Graham
Assistant Professor
Center for Learning and Memory
University of Texas at Austin

Shriram Ramanathan
Associate Professor
Applied Physics
Harvard University

PARTICIPANTS

Adriane B. Randolph
Assistant Professor of Information Systems, KSU BrainLab Director
Information Systems
Kennesaw State University

Parthasarathy Ranganathan
HP Fellow
Hewlett Packard Labs

Ravishankar Rao
Research Staff Member
Computational Biology Center
IBM Research

Karin A. Remington
Chief Technology Officer
Arjuna Solutions

Matt Richtel
New York Times

Anna W. Roe
Professor
Psychology
Vanderbilt University

Aviva Rutkin
Science Writing Scholar
Massachusetts Institute of Technology

Stephen Ryan
President, Doheny Eye Institute and Grace and Emery Beardsley Professor
Ophthalmology
Keck School of Medicine
University of Southern California

Paul Sajda
Professor
Biomedical Engineering
Columbia University

Terrence Sejnowski
Investigator, Howard Hughes Medical Institute; Francis Crick Professor
Salk Institute for Biological Studies

Thomas Serre
Assistant Professor
Cognitive Linguistic & Psychological Sciences / Brain Institute
Brown University

Kelly Servick
Science Writing Scholar
University of California, Santa Cruz

Rina Shaikh-Lesko
Science Writing Scholar
Science Communication Program
University of California, Santa Cruz

Sam R. Sharar
Professor
Anesthesiology and Pain Medicine
University of Washington School
 of Medicine

Jonathan Z. Simon
Associate Professor
Electrical & Computer
 Engineering / Biology
University of Maryland, College
 Park

Eliot R. Smith
Chancellor's Professor of
 Psychological and Brain
 Sciences
Department of Psychological and
 Brain Sciences
Indiana University

Daniel Stokols
Chancellor's Professor
Department of Planning, Policy,
 and Design and Department
 of Psychology and Social
 Behavior
Program in Public Health and
 Department of Epidemiology
University of California, Irvine

David L. Strayer
Professor of Psychology
Department of Psychology
University of Utah

Aaron D. Striegel
Associate Professor
Department of Computer Science
 and Engineering
University of Notre Dame

Kimberly A. Suda-Blake
Senior Program Director
National Academies Keck *Futures Initiative*

Diane M. Sullenberger
Executive Editor
PNAS
National Academy of Sciences

Laura L. Symonds
Assistant Professor of
 Neuroscience; Undergraduate
 Neuroscience Director
Neuroscience
Michigan State University

Mercedes Talley
Program Director
W.M. Keck Foundation

Desney S. Tan
Principal Reseacher
Computational User Experiences
Microsoft Research

Ashley Taylor
Science Writing Scholar
New York University

Kelly Tucker
Science Writing Scholar
MS Program in Science & Technology Journalism
Texas A&M University

Matthew A. Turk
Professor
Computer Science, Media Arts and Technology
University of California, Santa Barbara

Yalda T. Uhls
Researcher; Regional Director
Psychology
Children's Digital Media Center@LA
University of California, Los Angeles
Common Sense Media

Clara H. Vaughn
Science Writing Scholar
Philip Merrill College of Journalism
University of Maryland

Anthony D. Wagner
Professor of Psychology and Neuroscience
Psychology
Stanford University

Brian A. Wandell
Isaac and Madeline Stein Family Professor
Department of Psychology
Stanford University

Jun Wang
Research Assistant Professor
School of Information Studies
Syracuse University

Qi Wang
Assistant Professor
Department of Biomedical Engineering
Columbia University

Brian Waniewski
Managing Director
Institute of Play

Jason M. Watson
Associate Professor
Psychology
University of Utah

Debra L. Weiner
Assistant Professor
Emergency Medicine
Boston Children's Hospital/Harvard Medical School

Aaron L. Williams
Graduate Researcher
Systems and Information Engineering
University of Virginia

Kate Yandell
Science Writing Scholar
Arthur L. Carter Journalism Institute
New York University

Michelle Yeoman
Science Writing Scholar
MS Program in Science &
 Technology Journalism
Texas A&M University

Byron M. Yu
Assistant Professor
Electrical and Computer
 Engineering / Biomedical
 Engineering
Carnegie Mellon University

Tian Zhang
Fellow in Hematology-Oncology
Medicine
Duke University Hospital